Cases in Managerial Data Analysis

changing the way the world learns

To get extra value from this book for no additional cost, go to:

http://www.thomson.com/wadsworth.html

thomson.com is the World Wide Web site for Wadsworth/ITP
and is your direct source to dozens of on-line resources.
thomson.com helps you find out about supplements,
experiment with demonstration software, search for a job,
and send e-mail to many of our authors. You can even
preview new publications and exciting new technologies.

thomson.com: *It's where you'll find us in the future.*

Cases in Managerial Data Analysis

William L. Carlson
St. Olaf College

Duxbury Press
An Imprint of Wadsworth Publishing Company
I︎T︎P™ An International Thomson Publishing Company

Belmont • Albany • Bonn • Boston • Cincinnati • Detroit • London • Madrid • Melbourne
Mexico City • New York • Paris • San Francisco • Singapore • Tokyo • Toronto • Washington

Editor: *Curt Hinrichs*
Editorial Assistant: *Cynthia Mazow*
Marketing Manager: *Lauren Ward*
Print Buyer: *Barbara Britton*
Advertising Project Manager: *Joseph Jodar*
Permissions Editor: *Peggy Meehan*

Production: *The Book Company*
Design: *Beverly Baker, George Calmenson*
Cover Design: *Madeleine Budnick*
Copy Editor: *Steven Gray*
Compositor: *Bookends Typesetting*
Printing: *Malloy Lithographing, Inc.*

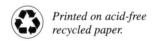

*Printed on acid-free
recycled paper.*

For more information, contact:
Wadsworth Publishing Company
10 Davis Drive
Belmont, California 94002, USA
USA

International Thomson Editores
Campos Eliseos 385, Piso 7
Col. Polanco
11560 México D.F. México

International Thomson Publishing Europe
Berkshire House 168-173
High Holborn
London WC1V 7AA
England

International Thomson Publishing GmbH
Königswinterer Strasse 418
53227 Bonn
Germany

Thomas Nelson Australia
102 Dodds Street
South Melbourne, 3205
Victoria, Australia

International Thomson Publishing Asia
221 Henderson Road
#05-10 Henderson Building
Singapore 0315

Nelson Canada
1120 Birchmount Road
Scarborough, Ontario
Canada M1K 5G4

International Thomson Publishing Japan
Hirakawacho Kyowa Building 3F
2-2-1 Hirakawacho
Chiyoda-ku, Tokyo 102
Japan

Printed in the United States of America

10   9   8   7   6   5   4   3   2   1

Library of Congress Cataloging-in-Publication Data

Carlson, William L. (William Lee), 1938
    Cases in managerial data analysis / William L. Carlson.
      p.   cm.
    ISBN 0-534-51721-8
    1. Management—Statistical methods—Case studies.   2. Problem
solving—Statistical methods—Case studies.   I. Title.
HD30.215.C37   1996
658.4'033—dc20   96-5521

To

Ruth and Leslie Carlson
for setting the goal!

Charlotte Carlson
for providing the support!

Andrea, Douglas, and Larry
Carlson
for the future!

PREFACE

This book provides the important link between the study of applied statistics and the capability to analyze problems in actual organizations. It offers real business applications, including the data used in those applications. The cases and data files are organized to emphasize problem solving and report preparation by managers and economists, without the overhead of data collection.

My motivation for writing this book is rooted in my teaching experience and in the modern consensus about teaching applied statistics. Good statistical theory and methodology must be combined with experience with real applications if you are to become a competent analyst. Initially I gathered data sets from actual projects that I had conducted. Later I obtained additional data sets from colleagues in business and industry. I developed these data sets into cases with a description of the setting, and students were asked to prepare an analysis and a report. Recently the annual conference on "Making Statistics More Effective in Schools of Business" has emphasized the importance of using applications and data for teaching statistical methodology and practice. Real business and economics applications of the type contained in this book go beyond the extended problems found in most statistics textbooks and provide a sound basis for problem solving.

UNIQUE FEATURES

This book includes the following important components:

- Based on real applications and data, the cases integrate issues from different functional areas, including management, finance, marketing, economics, quality assurance, and accounting.

- The cases contain large data sets and require computer software for analysis.

- Cases are of varying difficulty, permitting student growth through a number of case analyses.
- Guidelines and procedures for project analysis and project report writing are included.
- Teachers are given flexibility to custom-publish individual cases or subgroups of cases at a lower price.
- Extended solution materials are provided for teachers adopting the casebook.

All these components promote effective teaching of applied problem solving.

NATURE OF STATISTICAL STUDIES

Statistical studies generally proceed as follows:

1. Each statistical study begins with a clear statement of the problem situation and the rationale for selecting an appropriate analytical strategy.
2. This is followed by data collection, entry into computer files, and data checking.
3. Next, descriptive and analytical procedures are used to interpret the data and to provide answers to the analysis questions. In this process, efforts are typically devoted to correcting data, finding new insights, and (possibly) revising the initial questions.
4. Final conclusions are developed, based on the initial conclusions and analysis results.
5. Lastly, a written and/or oral report is prepared and presented for the client.

Experience teaches us that problem-solving processes are not linear: analysis results may lead to additional data checking; and writing the report often leads to additional questions and further analysis.

The most time-consuming tasks are data collection and entry, but in this casebook those tasks have been eliminated. This elimination certainly does not detract from the importance of good data collection and the lessons to be learned from actually collecting data. In fact, the task is arguably so important that it should not be presented as a burdensome addition to an already busy course work load. Instead, data collection should be part of an of extended research project that combines business functional areas and statistical analysis.

USING CASES IN STATISTICS COURSES

The Introduction of this book presents an approach to problem analysis and guidelines for preparing project reports. These are based on my experience using the cases in a variety of classroom situations. Initially you can follow the guidelines as presented; but I encourage you to expand and extend the basic framework and develop your own problem-solving methodology. Experience with these cases will provide a suitable environment for doing so.

Faculty who adopt this casebook will receive a data disk that provides easy interactive access to case data files in a format appropriate for most popular statistics packages and spreadsheets. In addition, the teacher's manual provides extended solution notes, all required computational output, and guidelines for assigning cases appropriate for your class and curriculum requirements. Given the wide range of business and economic functional areas included, teachers may decide to use some of these cases for joint projects with other courses in, for example, accounting, finance, marketing, or quality assurance.

Since each case can be completed in a realistic amount of time, the teacher can assign a number of different cases during the course. Typically I assign six or seven cases each semester. Generally the performance on the first case is not outstanding. By providing careful feedback in the grading process, however, I find that students noticeably improve during the semester. Good "statistical practice" is learned best through practice, scrimmage, or rehearsal.

ACKNOWLEDGMENTS

Many people contributed to the success of this project. Well over a thousand St. Olaf students have struggled with various forms of these cases. Their questions, criticisms, and quality project reports were basic to the formulation of the book. I obtained cooperation and assistance from numerous business people who are involved daily with real statistical problems. Their discussions of application situations and their actual data provided the material for the book. Included in this group are Anne Lundstrom, Robert Reul, Phillip Cartwright, Mark Hornung, Mike Giese, John Stull, Cliff Okerlund, Richard Niemic, Alice Miller, Mark Belles, James Donaghy, Keith Casson, Jean Bronk, James Munson, Jim Campbell, Pete Sibal, and Jack Litzau. Several faculty colleagues—Mary Emery, Rick Goedde, and Kathy Chadwick—provided valuable guidance in various business functional areas, and Linda Hunter provided valuable assistance in developing guidelines for the written reports. Douglas Scholz-Carlson prepared a number of TEX layout macros that greatly helped in the writing of this book.

Sincere appreciation is extended to a number of experienced faculty reviewers whose recommendations contributed significantly to this project. These reviewers whose recommendations contributed significantly to this project. These reviewers include Mohammed Askalani, Mankato State University; Michael S. Broida, Miami University; Michael Middleton, University of San Francisco, Paul Paschke, Oregon State University, Paul Randloph, Texas Tech University; Al Schainblatt, San Francisco State University; Paudu R. Tadikamalla, University of Pittsburgh; Stanley Taylor, California State University—Sacramento; Bruce Trumbo, California State University—Hayward; and Cindy Van Es, Cornell University.

Thanks to Scott Isenberg who initially proposed this project and developed the original contract. Editors Millicent Treloar and Curt Hinrichs provided valuable guidance and direction that brought this project to completion.

Finally I must acknowledge the loving support and encouragement of my wife Charlotte. Her willingness to put extra time into family activities during my days of intensive work on this book was a major contribution to the successful completion of it.

William L. Carlson
Northfield, Minnesota
January 1996
E-mail Carlson@stolaf.edu

CONTENTS

Cases in Managerial Data Analysis

Preparation of Case Solutions

Developing Statistical Skills

This casebook provides various typical business and government policy problems whose solution involves using statistical analysis. Business is seeking prospective employees who can carry out independent analysis and prepare completed project reports that solve real problems. Thus there is a clear need to move beyond statistics classes that only require solving textbook problems. Textbook problems are clearly important learning tools, but future business analysts need to learn how to examine the problem environment, prepare and implement a solution strategy, and write a clear, concise report. Over many years, participants in the annual conference on "Making Statistics More Effective in Schools of Business" have consistently endorsed this need. In addition to providing examples of real applications, these cases are an important motivator for learning statistical procedures; thus, when taught in the abstract, statistics can be confusing and boring. When used in the context of applied problems, however, it opens many doors and generates excitement as new insights are discovered in diverse areas of business. Applied statistical analysts need to link their statistical understanding to real problems.

The cases in this book reflect typical business applications. Most cases are based on descriptions of actual problems and on data received from a number of different firms and organizations. Descriptions of the organization and specific results have sometimes been disguised, however, to protect important confidential business knowledge. Typically, this involved modifying the original data by a constant or random factor. Such changes do not alter the business and management issues. In some instances, data contain outliers and measurement errors. Thus you must develop mature data-checking skills as you work on these cases. For that reason, the skills and experience gained in performing the analysis and preparing the reports will carry over to your future career.

Cases similar to these have been used for more than fifteen years in college classes. Most students view them as being the outstanding part of the course. Student evaluations indicate that, although the cases require considerable additional work, they make statistics much more interesting and greatly improve students' understanding of statistical procedures. Finally, students who perform well on the cases typically do well on subsequent exams.

CASE ANALYSIS PROCESS

These cases are designed to create a problem-solving environment that closely matches the environment experienced in an actual business. Case solutions require significant statistical analysis and interpretation of results. The focus is on creating a concise professional report that responds to the original business or public policy question. In contrast, textbook problems often require only a set of statistical analyses and computations. To simulate an actual business setting, you should use a microcomputer to perform the statistical computations and a text editor to prepare a professional-looking report. Working with a graphical interface such as a Macintosh or Microsoft Windows is ideal. These cases are perfect for teams of two students to tackle. Working in teams permits a broader examination of the case questions and fosters useful discussion. In addition, most real-life business problems are solved by teams. Thus students gain practical experience by working with others on a common problem.

Each case ends with a series of questions designed to help guide your analysis. The ultimate goal of your analysis, though, is a formal written report that helps solve the original problem.

Statistical studies follow a general structure, with a number of side diversions resulting from discoveries made along the way. The process usually consists of the following steps:

1. Identifying the analysis questions
2. Converting analysis questions into statistical questions
3. Performing a descriptive analysis of the data
4. Identifying unusual data points
5. Applying formal analysis procedures
6. Developing conclusions and recommendations
7. Identifying unusual outcomes and future analysis

Each of these steps should be accompanied by written descriptions in the final report text file. After all of the analysis and blocks of writing are completed, you may wish to move blocks of text and statistical output around to produce a more coherent report. The final step is then to prepare the Executive Summary, which is the very first section of your report. Each step entails a wide range of options.

1. IDENTIFYING THE ANALYSIS QUESTIONS

Your first task is to read the entire case carefully and make notes concerning key details. From this reading, you can begin to identify the important business questions in the case. In a real business application, this process consists of talking with key managers and other knowledgeable persons. That process usually leads to an analysis proposal, which is then "signed off" by the executive responsible for underwriting the study.

2. CONVERTING ANALYSIS QUESTIONS INTO STATISTICAL QUESTIONS

Statistical analysis can answer specific well-structured questions. You have seen many examples of these questions in textbook problems and examples. An important task in any applied study is to convert the case analysis questions into statistical questions. This process requires thorough understanding of the problem requirements and of the available data. In many cases the available data, including the data form and source, limit the kinds of analysis that are possible.

3. PERFORMING A DESCRIPTIVE ANALYSIS OF THE DATA

Performing a descriptive analysis provides the analyst with a general picture of the data. Means, variances, ranges, and simple correlations are computed for appropriate variables. The analyst may also prepare histograms, box-and-whisker plots, and other graphical displays. If the data have been collected over time, creating time plots will help identify patterns that change over time. From this analysis, you can define the range of applicability for your analysis conclusions.

4. IDENTIFYING UNUSUAL DATA POINTS

Data from real business applications frequently contain some unusual observations or outliers. In many cases these result from simple recording or measurement errors. In other cases, however, strange points may indicate important observations of the environment that generated the data. In all cases each unusual data point should be traced to its source, and the error should be corrected or the special condition reported. For symmetrical distributions, about 95 percent of the data set is contained within plus or minus two standard deviations of the mean and very few observations lie outside plus or minus three standard deviations. This rule of thumb can help identify extreme points. Similarly, points that are far from a central linear grouping on a scatter plot should be investigated. Detecting strange data requires knowing how to use the statistical tools in your computer package and being interested in problem solving.

Several statistical procedures can be employed to deal with outliers. Usually they require some assumptions about the data generation process and the underlying probability model. These should be used if you understand their assumptions and if the procedure is appropriate. Analysis quality can be improved by using sophisticated analysis procedures properly. Applied work raises two fundamental questions about outliers: (1) is the data outlier merely a recording error? (2) did the observation actually occur? If the answer to the first question is yes, the error should be corrected. If the answer to the second question is yes, the observation must be included in the analysis and report. But appropriate adjusting procedures can still be used.

5. APPLYING FORMAL ANALYSIS PROCEDURES

By this point in the study, you will have developed a good understanding of the problem issues and the basic structure of the data. In addition, if you have been writing intermediate results, you will probably have additional questions. These can be compared against the assumptions required for the various standard statistical procedures. Do you have everything needed for a formal hypothesis test? Do the data satisfy the assumptions required for regression analysis? Which of the available analyses will answer the questions you have formulated?

The analysis should be driven by the questions and not by a desire to demonstrate how clever you can be with various statistical tools. You should be willing to try a variety of procedures to test the stability of the conclusions. For example, means, standard deviations, and other descriptive statistics should show the same results as bar charts and graphs; and regression models of different forms should provide similar conclusions. Analysis of residuals and analysis of outliers in general evince a careful analyst. Study of the distribution patterns for important variables indicates the validity of your analysis. Many strange patterns can occur in data, and you should try the analysis in enough different ways to ensure reliability of the results. Incidentally, the ability to support different analyses is a major advantage of the computer analysis environment.

This is also the time to perform some basic reality checks on your conclusions. Do the results make sense, given what you know about the area being studied? You can also ask other persons with experience in problems of this type to react to your initial conclusions.

Computer-based problem solving can create undeserved support for certain answers, if there were initial errors in the analysis. Older, more tedious non-computer-based analysis procedures were closer to the data and the detailed steps. Thus, errors in computation or analysis approach were more likely to be detected. In our modern computer-based statistical analysis environment, results must be examined for reasonableness and reality.

6. DEVELOPING CONCLUSIONS AND RECOMMENDATIONS

At this point you should review the initial objectives and questions to confirm that all of the required analyses have been completed. If you have been writing after completing each analysis step, you already have the basis for discussion of your major conclusions. Key ideas should be extracted and combined for a smooth-flowing discussion. Some of the more basic details can be moved to an appendix. This is also the time to write the Executive Summary.

Ideally this step involves bringing together your results and then writing a clear discussion that links them. However, real studies do not always proceed linearly (such that the previous step is completed before the next step is begun). Thus you may identify additional questions when you are preparing your conclusions. In that case you must cycle back to perform additional analysis.

7. IDENTIFYING UNUSUAL OUTCOMES AND FUTURE ANALYSIS

Research, analysis, and problem solving are an ongoing process. Answers at one point identify additional questions. Analysis often indicates additional paths that should be followed. However, studies in a business or public policy environment must reach a close. Decision makers require timely answers, and limited resources are allocated to each study. Personnel and financial resources are allocated to a study, and results must be produced within resource limits. As a student, you have only so many hours to devote to each course and/or to special projects. Thus completion of a study is important. Closure typically includes a final written report and/or an oral presentation.

Given the need for closure, you will often be left with unexplored questions in your study. These should be documented and noted as questions for future work. In some cases the questions might be important enough to justify another study. Your task is to document the additional questions objectively and to suggest how they might be answered. Information users must then decide whether additional work should be undertaken.

PREPARING THE WRITTEN REPORT

Since producing a written report is your ultimate objective, you would be wise to begin writing your report after you have identified the analysis questions but simultaneously with performing the statistical calculations. For example, start the case report by preparing the introduction and the project scope, using your text editor. Initial statistical analysis can then be saved in a file and transferred to your emerging report. Next, prepare comments on the initial statistical work, and move back to perform more statistical computations. This strategy has the advantage of encouraging you to keep your results well organized as you proceed. More importantly, the process of writing forces you to define specific questions that you can then answer by using statistical analysis. And good writing almost always requires several revisions.

For these cases, your report will typically be limited to two or three pages, with extended appendices that present specific analyses and data displays. The report should begin with an Executive Summary of one to three paragraphs. This summary—which is the last item written—should identify the problem, indicate your approach to solving it, and concisely state your conclusion.

The body of your report should indicate how you developed your conclusions and recommendations. Begin with a concise presentation of the question from the business perspective and explain how you conducted the analysis. Define the data set and specify the statistical procedures you used. Include specific statements of statistical models and hypothesis tests, and outline the results. Discuss the statistical results, indicating how they provide a solution to the case problem. Include additional observations and extensions of your results, as appropriate, given the case objective.

Business writing tends to be short and concise. Managers and executives seek results and conclusions. They need to understand your analysis and conclusions. In addition, any limitations and alternatives need to be clearly expressed. Managers do not have time for extensive reading of analysis details. In many instances, the Executive Summary is the only part of a report that is read. Thus you should prepare the Executive Summary with great care. The results of all of your hard work may be revealed or hidden depending on the quality of the Executive Summary.

Persons who seek more detailed information will read the entire report. The report should describe your analysis and present conclusions together with supporting evidence. Well-designed graphs and figures greatly enhance communication. Mathematical equations and data tables provide strong reinforcement for readers who desire rigorous and complete understanding. However, you need to avoid both intimidation and boredom. Analysis details, including detailed statistical computer outputs and supporting graphs, should be in an appendix. These reports are a series of layers that supply further depth as more accessible layers are peeled away. The first layers, however, must provide the most important results and key conclusions. Think about this design for your report as you prepare case solutions.

CONCLUSION

Work on the cases in this book can provide a valuable capstone to your study of applied statistics. By developing solutions to applied problems you will think carefully about the assumptions behind various statistical procedures.

TABLE 1.1	LIST OF CASES

CASE NUMBER	CASE NAME	DATA FILE	FUNCTIONAL AREA
1	Allflex Inc.	Allflex	Quality Control
2	Olson Diversified Marketing Services	Olson	Accounting
3	Stardoe Coffee Inc.	Stardoe	Marketing
4	Fort Worth Bay Bagels Ltd.	Bagel	Business Operations
5	National Cotton Fabric Association	Cotton	Marketing
6	Sports Unlimited Inc.	Sports	Accounting
7	Consolidated Foods Inc. A	Confood	Marketing
8	Consolidated Foods Inc. B	Confood	Marketing
9	Consolidated Foods Inc. C	Confood	Marketing
10	Consolidated Foods Inc. D	Confood	Marketing
11	Midwest Equipment Inc. A	Midwest	Accounting
12	Midwest Equipment Inc. B	Midwest	Accounting
13	Midwest Equipment Inc. C	Midwest	Accounting
14	Prairie Flower Cereal Inc. A	PrairieA	Manufacturing Operations
15	Prairie Flower Cereal Inc. B	PrairieB	Manufacturing Operations
16	Prairie Flower Cereal Inc. C	PrairieC	Quality and Process Control
17	Prairie Flower Cereal Inc. D	PrairieD	Quality and Process Control
18	River Valley Insurance Co. A	Health	Health-care Management
19	River Valley Insurance Co. B	Health	Health-care Management
20	State Lottery A	Lottery	Government Business Operations
21	State Lottery B	Lottery	Government Business Operations
22	National Crime Control Program	Crime	Government Social Program
23	American Motors Inc.	Motors	Business Planning
24	State Planning for Loonland	State	Government Fiscal Policy
25	Production Systems Inc.	Prodsys	Human Resources
26	Sheldahl Inc.	Sheldahl	Quality and Process Control
27	New Concepts Financial Inc.	Concepts	Portfolio Risk and Return
28	Big Sky Power Inc.	Bigskyrg, Bigskyht	Usage Forecasting

This will broaden and deepen your understanding of statistics. By working on business and public policy problems, you can see statistics as an important part of your future career and not just a difficult supporting discipline required for your major. Applied statistics is exciting because it helps lead to new insights. The most exciting applications of statistics, however, occur in an interdisciplinary problem-solving context. The cases in this book provide some of that context. Table I.1 presents a list of these cases, along with the data file name and the functional application area of each.

ALLFLEX INC.

Production and Quality Performance Analysis

1.1 INTRODUCTION

Allflex Inc. is a new entrepreneurial business that manufactures flexible circuits in a small midwestern city. The firm was started by Anne Lundstrom, who is president and chief operating officer, to serve a specialized small-market niche. The company started with three employees and in two years has grown to twenty employees.

Based on her fifteen years of engineering experience in the electronics industry Anne concluded that there was a continuing market for a company that could produce small quantities of flexible circuits (orders of less than 300) with a short lead time (less than 25 working days from order to delivery). Manufacturers of specialized electronic equipment and prototypes are constantly in need of small quantities of specialized connectors to link the various electronic components in their equipment. In many cases such manufacturers are under tight delivery deadlines, and design changes often occur at intermediate stages of the project. Flexible circuits for connecting components typically account for only a small part of the total equipment project budget. They are critical, however, and procurement delays or low-quality circuits can lead to major production delays and operating problems in the resulting equipment. Thus the price per unit is not of major concern to the buyer, but timely delivery of a high-quality product is.

Flexible circuits contain a number of electrical connector paths embedded in a flexible plastic resin material. They are used to connect various electronic components such as computer processors, memory chips, and control devices. With the explosion of microprocessor-based controls in a wide range of products and equipment, flexible circuits have become a key component in automobiles, computers, robots, medical equipment, and other complicated machinery.

Flexible circuits are produced from an original sheet of thin copper bonded to a flexible resin material. The circuit design layout is typically prepared on a computer-assisted design (CAD) machine, with lines on a drawing representing the connection paths. Part of the designer's task is to obtain a maximum number of circuit units from each sheet of material used in the process. Since manufacturing costs are directly related to the number of sheets processed, costs per unit are reduced if more units can be produced from a single sheet. However, crowding units or placing electrical paths too close together on each circuit can create quality problems. A final drawing showing the layout of circuits on a standard sheet is prepared; this represents the basic manufacturing unit. The circuit drawing is photographed and then etched on the copper material. Excess copper, which is not part of the connection paths, is removed by passing the sheets through various acid baths.

CRT Technologies, a manufacturer of robotics and controls, is seeking a long-term supplier for the many different flexible circuits used in the equipment they manufacture. As a subcontractor, CRT produces equipment for new factory or manufacturing installations within existing plants. Typically CRT must provide specialized equipment design and manufacture or extensive modifications of previous designs. CRT has developed a reputation for completing contracts on time with a minimum of problems associated with equipment design and manufacture.

Flexible circuit connectors are far from the most expensive components of CRT's custom-made equipment. However, connector failure can greatly increase installation time and cost, and it can damage the company's reputation for producing high-quality, trouble-free manufacturing systems.

Allflex has prepared a proposal to become sole supplier of flexible circuit connectors to CRT under a cost-plus contract. Prior to entering into such a contract CRT wishes to be assured that Allflex has sufficient orders to maintain itself as a strong firm over an extended period. In addition, CRT is seeking evidence that Allflex is improving the quality of its products. Allflex has submitted the data in Tables 1.1 and 1.2 and Exhibit 1.1, as well as in a spreadsheet data file titled Allflex. Table 1.1 presents figures on total production and number of defects per month, measured in sheets. Table 1.2 and Exhibit 1.1 present a breakdown of defects by various manufacturing stations. The computer spreadsheet contains these data in a form suitable for statistical analysis.

You have been asked to analyze this information and prepare a report for CRT concerning the viability of Allflex as a long-term supplier of flexible circuits. Your report should examine the available data, indicate the production and quality performance of Allflex for the past nine months, and offer a recommendation based on your prognosis of Allflex's future performance.

Your analysis should include the following steps.

1.1 Load the spreadsheet file named Allflex into your local computer system.

1.2 Analyze the time-based changes in total output and in number of defects.

1.3 Examine and analyze the detailed description of defects by production area.

1.4 Prepare a written report summarizing your conclusions and presenting a recommendation concerning the viability of Allflex Inc. as a supplier for CRT.

TABLE 1.1	MONTHLY PRODUCTION AND DEFECTS	
MONTH	**SHEETS PROCESSED**	**SHEETS SCRAPPED**
October	571	110.73
November	665	84.70
December	891	116.45
January	1077	96.10
February	1472	54.68
March	1713	168.30
April	1300	92.62
May	1399	121.31
June	2136	64.10

TABLE 1.2	DEFECT FREQUENCY BY STATION											
MONTH	**A**	**B**	**C**	**D**	**E**	**F**	**G**	**H**	**I**	**J**	**K**	**OS**
October	0.00	0.07	3.19	5.08	5.06	51.79	21.86	0.67	1.42	0.00	—	21.58
November	0.88	1.50	2.51	0.00	0.00	24.77	43.05	6.31	2.13	0.00	—	3.56
December	0.57	0.03	1.92	4.82	0.00	62.39	39.90	4.18	2.20	0.00	—	0.44
January	5.66	0.00	1.39	2.65	0.00	25.30	31.70	0.78	5.52	0.11	4.00	19.00
February	0.00	0.31	0.50	4.21	0.31	14.90	30.29	0.54	1.15	0.00	0.00	2.47
March	1.31	18.64	26.94	31.47	0.00	22.75	42.53	0.79	9.06	0.27	14.59	2.50
April	0.58	0.00	4.12	0.20	7.79	35.27	26.57	5.34	2.43	7.00	0.14	2.58
May	0.41	0.00	1.08	0.91	0.11	18.43	89.03	3.65	6.00	0.00	0.11	1.60
June	0.00	0.04	0.60	5.77	1.17	8.27	45.01	0.08	0.08	0.00	3.00	0.08

EXHIBIT 1.1	DEFECT CODES BY STATION

Material Preparation Station

A1 Error on Route Sheet
A2 Incorrect Material Selected
A3 Damaged Material (Scratched, Dented, Bent)
A4 Other

Material Drilling

B1 Drilling Damage
B2 Incorrect Hole Size Drilled
B3 Missing/Extra Holes
B4 Wrong Program Used
B5 Drill Registration
B6 Other

First Copper

C1 Voids
C2 Adhesion
C3 Copper Not Thick Enough
C4 Dark or Grainy Deposits
C5 Extra Copper
C6 Other

Imaging

D1 Delamination of Resistance Material
D2 Image Registration
D3 Annular Ring
D4 Dimensional Stability
D5 Handling Damage
D6 Scratches or Dirt on Artwork
D7 Underdeveloped Materials
D8 Other

Electroplating

E1 Plating Voids
E2 Copper-to-Copper Adhesion
E3 Lifting of Resistance
E4 Burned Copper
E5 Plating Thickness
E6 Other
E9 Incomplete Resistance Stripping

Etching

F1 Incorrect Line Width and Spacing
F2 Etch-outs
F3 Damaged by Etcher

F4   Tin Not Removed
F5   Lifting of Resistance
F6   Overetching
F7   Incomplete Etching
F8   Other

In-Process Quality Inspection

G1   Electrical Short in Circuit
G2   Open Circuit
G3   Incorrect Hole Size
G4   Handling Damage
G5   Foreign Material
G6   Missing/Unknown
G7   Other

Cover Lay Station

H1   Clearance Hole Diameter
H2   Registration
H3   Annular ring
H4   Other

Die Cut

I1   Delamination Caused by Die Cut
I2   Blanking Damage (Rough Edges, Torn, Dents)
I3   Dimensions Incorrect
I4   Scratches
I5   Other

Final Inspection

J1   Fail Solderability Test
J2   Fail Dimensional Test
J3   Fail Solder Mask Cure/Adhesion
J4   Fail Folding Flexibility Test
J5   Fail Thermal Stress Test
J6   Other

Assembly Station

K1   Assembly Error

Outside Services

OS1   Missing Parts
OS2   Shipping Damage
OS3   Defect from Leveling
OS4   Defect from Small Adhesion
OS5   Defect from Other Outside Service

OLSON DIVERSIFIED MARKETING SERVICES

Analysis of Receivables

2.1 BACKGROUND

Olson Diversified Marketing Services[1] is a medium-sized business service organization that provides sales and promotion services to various medium- to large-size companies, including a number of Fortune 500 companies. Olson's services include publicity and redemption of coupon-based marketing campaigns, sampling of trial sizes of new products to potential customers, management and administration of product promotion contests, administration of sales-force incentive programs, and planning and operation of national sales meetings for major corporations. For every project contracted, Olson assigns a project manager and various staff professionals. In addition, the company may hire consultants, performers, and diverse services, as required by the contract. When the project is completed, Olson prepares a complete billing, including overhead and a management service fee. After the bill has been received, the client is required to submit payment within 15 days or pay contracted interest payments. Olson Diversified Marketing Services has established a target of 45 days after the end of the project to complete its account of accumulated charges and to submit a bill to the client.

The most recent audit by outside auditors has revealed that accounts receivable (money owed to Olson by clients) has grown considerably. Auditors have demonstrated that the receivables constitute too large a percentage of Olson's sales and have advised the company to develop a plan to substantially reduce receivables. Company management believes that reducing the fifteen-day period for client payment would seriously damage its relationships with present and future clients. Thus any improvement in the

[1]Acknowledgment and appreciation are extended to Carlson Marketing Group of Minneapolis and to Robert Reul, managing director of quality systems and methods, for supplying the data and problem background.

receivables problem must result from reducing the time required to prepare clients' bills.

You have been assigned to analyze the problem and recommend a procedure to substantially reduce the receivables. First, you are to study the pattern of times currently required to prepare clients' bills and the bill amounts. Next, you must determine the potential savings from reducing the number of days spent preparing the bills. Prior to beginning your analysis, you meet with George Reale, manager of billing and receivables, in the accounting department. He has been concerned about this problem for some time, but heretofore his concerns have not been given serious consideration. Thus he is pleased that you are conducting the study and offers to help in any way he can. He provides you with a data disk containing information on the 445 projects completed and billed during the previous calendar year.

To reduce the time needed to prepare clients' bills, George has suggested implementing a new computerized billing system. This system would require project managers to enter all purchase orders and staff hours as soon as the services are completed. Suppliers would be asked to submit their charges electronically, and these charges would be added to the charges data file as soon as received. At the end of the project, a list of outstanding charges from suppliers would be prepared, and the project manager would direct an effort to obtain all remaining supplier charges. Previously, senior management argued that developing such a system would be too expensive and that operating it would necessitate too much extra work. After your meeting, you and George agree that you will focus on the potential cost savings and then determine whether these are sufficient to justify the computerized system proposed by George.

2.2 TECHNICAL BACKGROUND

Accounts receivable are funds owed by clients for goods or services provided. In accounting terms, accounts receivable are recognized when the services are performed. In addition to the actual charges, the receivables include a markup to cover profit, management, and allocated overhead. Receivables impose a cost on Olson because the company must borrow funds to cover payments to its suppliers before payment is received from the client. Olson maintains a line of credit with a local bank at an annual interest rate of 10 percent to cover these receivables. Thus, every $1,000 dollars held as receivables for the year results in an interest charge of $100. If Olson can reduce the time for a $1,000,000 client charge by one day, the net savings is $274 [(0.10)($1,000,000)/365]), and a savings of ten days is worth $2,740. Olson regularly manages a number of contracts, and the total amount of funds awaiting client payment is sometimes in the tens of millions of dollars. The amount of savings obtainable by reducing the billing time for each client contract is equal to the dollar amount of the client bill multiplied by the 0.10 bank interest rate, with the result then multiplied by the reduction in billing time (in days) divided by 365.

2.3 DEFINITION OF ANALYSIS PROJECT

You have been asked to study the pattern of bill preparation times cross-referenced by the size of client bills. From this research, you are to indicate the potential savings that could result from reducing billing time. You might, for example, recommend that efforts be concentrated on bills for amounts above a certain level. Or you might recommend that efforts be applied uniformly to all bills. Generally, larger bills entail substantially more cost items and thus necessitate more work to accumulate all charges.

The data file supplied by George Reale contains 445 observations on two variables. The first column indicates the number of days required to prepare and submit the bill to the client, and the second column indicates the total charges billed for the project. The data are stored in a data file named Olson. Your case project will include the following tasks.

2.1 Load the data from the file named Olson into your local computer system, and name the variables.

2.2 Prepare frequency distributions and histograms for the number of days for bill preparation and for the dollar value of the client's bill. Describe in words the distributions of these two variables.

2.3 Compute the descriptive statistics for bill preparation time and for the amount of the bill. Include the median and the upper and lower quartiles in the descriptive statistics.
  a. What percentage of the time does the company reach its goal of preparing the bill in 45 days?
  b. Describe the distribution of the number of days for bill preparation.
  c. Describe the distribution of the size of the client bills.

2.4 Determine the relationship between number of days and billable amount in two ways.
  a. Prepare a scatter plot.
  b. Use least squares regression to develop a model that describes the relationship between the billable amount and the number of days outstanding.[2]

---

[2]In this case use the computer's simple regression package to compute the constant and the slope coefficient for the linear equation that describes the relationship. You do not need to consider any other statistical information in the computer output.

2.5 Consider the 25 percent of all bills that have the highest charges.
  a. What is the average value of these bills?
  b. What is the average number of days required for preparation of these bills?
  c. How much money could be saved if the average bill preparation time could be reduced by 20 percent for these bills?

2.6 Repeat the preceding analysis, considering in turn the bills with the highest 50 percent, 40 percent, 30 percent, 20 percent, and 10 percent of charges.

2.7 Based on your analysis in parts 2.4, 2.5, and 2.6, recommend a strategy for selecting the number of bills that should receive extra effort to reduce the billing time.

2.8 Prepare a two-page written report that describes your analysis and develops a recommendation for reducing the cost of receivables.

STARDOE COFFEE INC.

Expansion into a New Region

3.1 INTRODUCTION AND COMPANY BACKGROUND

Oscar Lee, marketing vice-president of Stardoe Coffee Inc., has asked you to assist in the analysis of their proposed expansion into new regional market areas. First, he wants you to estimate the market potential in a specific proposed region. You are to develop a random sampling study and analyze the results. Then you must prepare a report recommending a course of action regarding Stardoe's proposed market expansion. Oscar has indicated that he would like to use your initial analysis as a format for analyzing other proposed expansions. After completing the analysis, you should recommend changes to the analysis procedure that would improve future studies.

Stardoe Coffee is a producer and distributor of high-quality gourmet coffee beans. The company was started in Seattle fifteen years ago by Mavis and Albert Johnson.[1] Mavis and Albert developed a taste for gourmet coffee during their ten years with the United States Foreign Service diplomatic staff. After moving to Seattle they experienced considerable difficulty in obtaining high-quality coffee. Initially Mavis would have coffee shipped from suppliers in various foreign countries whom they had met during overseas assignments. Their friends also enjoyed these coffees and asked Mavis to order additional coffee for their use. Finally Mavis and Albert decided to open their own business; importing and roasting coffee and distributing it in small coffee stores.

The couple devoted considerable effort to their business in the beginning. They hired an expert coffee roaster to develop and manage the production operation. Albert spent considerable time developing the process of brewing coffee for direct sale in retail stores and for storing beans to ensure freshness and high quality. Mavis developed a product image, designed the stores, and

---

[1]This case has no connection with any existing business. It is simply designed to present a realistic problem-solving environment. Any similarities between the Johnsons and actual persons are purely coincidental.

negotiated contracts for coffee purchase. As a result, their business grew from one store to twenty-five in the greater Seattle area. Annual sales increased by 15 percent each year until the past two years. Thus they concluded that the local market for their product was saturated.

Mavis and Albert believe that expansion into other urban markets offers good growth possibilities. This belief is based on the many positive comments they have received from customers visiting Seattle. In addition they have developed a modest national mail-order distribution, in spite of the very high cost per pound of coffee sold via this distribution channel. The marketing vice president, Oscar Lee, was assigned the task of preparing a feasibility study for possible expansion into new markets.

3.2 COST ANALYSIS FOR EXPANSION

Before talking to you, Oscar undertook an accounting study to ascertain the expected costs of establishing distribution in a medium-size urban area containing 350,000 households. The proposed design included a regional office and five retail stores. The regional office would require about $30,000 to establish and $130,000 annually for a regional manager, an administrative assistant, rent, utilities, and other fixed expenses. The costs for each store per month were estimated to be as follows: rent, $1,500; insurance, $300; utilities, $250; staff, $2,500; and other, $600. Each store would require an initial investment of $50,000 in fixed setup costs. The retail price is $8.00 per pound, and the cost of goods sold is $5.00 per pound, which leaves $3.00 per pound to cover distribution, store operations, and profit. Past experience indicates that shipping from the Seattle production facility would cost $0.10 per pound.

The costs provided in the accounting report are to be used to determine the sales volume required to provide an economic justification for the expansion. Experience indicates that households that respond positively to this coffee will purchase an average of 5 pounds per year. The company policy for expansion is that there must be strong evidence that, during the first year, revenues will cover all operating costs plus half of the initial setup costs.[2] Your first task is to determine the percentage and the actual number of households that must become customers if Stardoe is to achieve its first-year sales objective.

3.3 MARKET SURVEY DESIGN

You met with the marketing staff to design the sampling survey. After some discussion the group decided to use a short telephone interview. A company

---

[2]For purposes of this case, "strong evidence" means a probability of less than 0.05 of being wrong.

TABLE 3.1		VARIABLE NAMES IN THE STARDOE DATA FILE	
VARIABLE NUMBER	**VARIABLE NAME**	**NUMBER OF OBSERVATIONS**	**VARIABLE DESCRIPTION**
1	Intnumb	900	Identifying number for interview
2	Age	900	Age of principal wage earner
3	Income	900	Total household income
4	Educatin	900	Highest education level for principal wage earner 1 = Less than high school 2 = High school graduate 3 = 1 to 4 years of college 4 = Bachelors degree 5 = Graduate degree
5	Numhshld	900	Number in household (max. 5)
6	Gcoffee	900	Regular user of gourmet coffee (0 no, 1 yes)
7	Starcofe	900	Purchaser of Stardoe coffee 1 = yes 2 = maybe 3 = no

skilled in obtaining random samples and conducting phone interviews was selected to administer the survey. The questionnaire solicited data on four important demographic items: age, income, and education level of the person with the largest income; and the number of people in the household. Respondents were also asked if members of their household consume at least 5 pounds of gourmet coffee per year. Persons who responded positively to that question were asked if they would purchase Stardoe coffee beans. If they were not familiar with Stardoe coffee, a short description of the product was read to them. The responses to this last question will be used to estimate the number of households and the proportion of the market that will buy Stardoe coffee.

Your next task is to determine the required sample size. Data from other cities indicates that 10 percent of the urban population in those cities would purchase Stardoe coffee. The required sample size is to be based on a 95 percent two-tailed acceptance interval. Compute the sample size required to obtain this objective. Compare this sample size, based on the company objectives, with the sample size actually used when the survey was made

Table 3.1 contains the names and descriptions of the variables collected in the survey. These data are stored in a file named Stardoe. The data in the file were obtained by Market Knowledge Inc., which used a sample of randomly selected telephone numbers to select the subjects. Using these data, you are to determine whether there is sufficient market demand to meet the company objectives for expansion. In addition, the marketing department has asked you to identify the market segments that are most likely to purchase Stardoe gourmet coffee. Experience in the Seattle market suggests that preference is directly linked to income and education.

After conducting your series of discussions with various staff and your own analysis, you meet again with Oscar Lee to work out the specific requirements for the study. Based on this you agree to perform the following tasks.

3.1 Compute the number of households and the percentage of the 350,000 households needed to meet Stardoe's expansion criteria. Your analysis will use the cost and revenue data in this case.

3.2 Prepare descriptive statistics and graphical displays indicating the basic patterns for each of the variables in the sampling survey.

3.3 Compute the required sample size if the proportion is around 0.10 and if the 95 percent confidence interval is to be ±0.02.

3.4 Prepare a hypothesis test to determine whether strong evidence exists to support the conclusion that there is sufficient demand to justify expansion into the new urban market. Use the data in the sample data file to carry out the hypothesis test.

3.5 Determine the market segments that are likely to be the largest consumers of Stardoe gourmet coffee.

3.6 Prepare a written report for Oscar Lee that presents the results of your analysis and makes a recommendation about the proposed new market. The report should also include recommendations regarding which market segments to emphasize when selecting store locations or choosing promotion strategies.

FORT WORTH BAY BAGELS LTD.

Customer Arrivals at a Small Shop

4.1 INTRODUCTION AND BACKGROUND

Arthur Eckstein has been operating a successful bagel shop in the commercial district of a large metropolitan area. Recently, however, he has been receiving complaints from his regular customers, who believe that their waiting time for service has increased; indeed, some have begun patronizing other shops for their quick lunch or snack. Arthur has asked you to develop staffing recommendations for various times during the day. He has collected data on the number of customers per hour for the past 20 weeks. The data, which are stored in a file named Bagel, are described in Table 4.1.

Arthur established Forth Worth Bay Bagels Ltd. approximately three years ago. Previously had worked for several different companies that franchised small restaurants. In college he had worked for a very successful bagel bakery that sold hot bagels to students and faculty near the edge of campus. During that time he developed excellent skills for producing and marketing bagels. His restaurant experience provided a wide range of relevant experience. However, he often thought about the possibility of owning his own small restaurant some day. Finally, three years ago, he decided to pursue his dream and opened his own bagel shop.

Fort Worth Bay Bagels prepares bagels daily in its bakery. Bagels are brought to the downtown shop and are also distributed through selected neighborhood food stores, supermarkets, and convenience stores. This case involves only the downtown shop, which sells bagels for takeout and prepares many types of bagel sandwiches on request as customers arrive. A large group of people come to the shop for a quick lunch or snack during their business day. In addition, people will sometimes conduct short informal meetings over coffee and a sandwich. Generally the shop has become an extension of the business offices—especially for people in finance, real estate, and banking. Thus, there is an atmosphere of informal professionalism among the customers. Men and women customers are focused on their work and enjoy the atmosphere, with light food serving as a background for their work. Given this environment, customers seek a fresh, high-quality product and quick

TABLE 4.1		VARIABLE NAMES IN THE BAGEL DATA FILE
VARIABLE NAME	**NUMBER OF OBSERVATIONS**	**VARIABLE DESCRIPTION**
Day	100	Code identifying the sequential day of observation
Time7	100	Number of customers, 7 A.M. to 8 A.M.
Time8	100	Number of customers, 8 A.M. to 9 A.M.
Time9	100	Number of customers, 9 A.M. to 10 P.M.
Time10	100	Number of customers, 10 A.M. to 11 A.M.
Time11	100	Number of customers, 11 A.M. to 12 noon
Time12	100	Number of customers, 12 noon to 1 P.M.
Time1	100	Number of customers, 1 P.M. to 2 P.M.
Time2	100	Number of customers, 2 P.M. to 3 P.M.
Time3	100	Number of customers, 3 P.M. to 4 P.M.
Time4	100	Number of customers, 4 P.M. to 5 P.M.
Time5	100	Number of customers, 5 P.M. to 6 P.M.
Time6	100	Number of customers, 6 P.M. to 7 P.M.

service. Waiting for service is viewed as a waste of their busy work time and is generally annoying to customers. Arthur understood his market well when he opened his shop. He created a high-quality product and emphasized rapid service. Business people learned of the shop, and many became regular customers.

As the business has grown, delays have recently become a critical problem. When customers enter the store, they go to a central counter where they can order a sack of bagels, drinks, or bagel sandwiches. Serving a customer usually takes about three minutes. Several servers are stationed behind the counter, and they handle customers as quickly as possible. The number of servers varies depending on the anticipated number of customers during that hour. If all servers are busy, all additional customers must wait. Typically, however, the order of the next person in line is taken within three minutes.

Arthur charges premium prices and believes strongly that he should have sufficient staff available so that most customers do not have to wait for service. Customers usually arrive individually, and the number who arrive during each three-minute period can be modeled by the Poisson probability distribution. After some discussion with Arthur, you agree that there should be enough servers to ensure that approximately 80 percent of all customers will not have to wait when they enter the shop.

The data file named Bagel contains the number of customers per hour for the past 20 weeks. The shop is open Monday through Friday, so there are observations from 100 weekdays. The shop is open from 7:00 in the morning until 7:00 in the evening. Arthur indicates that his policy is to hire servers for at least three consecutive hours during a day, and not all workers work every day. However, he prefers to employ servers full time—that is, eight consecutive hours daily, five days a week. He believes that regular employees provide better service because they have a direct interest in the success of the business. In addition, customers develop a personal relationship with servers, and this encourages the customers to return.

4.2 TECHNICAL NOTE

This case presents a classic problem for businesses that serve large numbers of customers as they arrive. Similar situations occur at bank teller windows, airline check-in counters, highway toll booths, and telephone systems. Based on the expected number of arrivals, the company schedules a certain number of units to provide service to arriving customers. In this case Arthur Eckstein schedules a number of servers to prepare customer orders. If customers arrived at a uniform rate the scheduling task would be easy. For example, if one customer arrived every minute and service required three minutes, the firm could schedule three servers; then no customer would ever need to wait, and the workers would be busy all of the time. But, in the real world, customers tend to arrive at very short intervals for a while, followed by periods of few arrivals. Generally, the precise pattern cannot be predicted.

These problems are classified as queuing theory problems, and a number of special models have been developed for their analysis. Many of these models are highly complex and require considerable graduate study in probability and stochastic processes. However, some problems can be handled using the Poisson probability model that you have studied. The basic assumption here is that people arrive independently at a predictable average rate for a specific time period. In this case we can assume that the arrival of one person does not affect the arrival of another person. Each customer decides at some point to interrupt his or her work and have a quick lunch. Of course, the average number will be higher during certain time periods (such as the usual lunch period or the breakfast period at the beginning of the day) than at others.

For this case we know that a customer is served on average in three minutes. Thus the number of servers required for any hour should be such that the probability that all customers who arrive during a three-minute interval will be served at once is at least 0.80. For example, if five servers are scheduled, the probability of five customers arriving should be less than or equal to 80 percent and the probability of six customers arriving is greater than 80 percent. In this example, customers would have to wait only if more than five arrived during a three-minute period. The decision to schedule five servers would have been made only if the probability of more than five customers arriving was less than 20 percent (1– 0.80). By choosing the number of servers in this way, we say that the "customer service level" (CSL) is 80 percent; that is, at least 80 percent of the customers will be served without having to wait. From the data collected, the mean number of arrivals for each hour can be computed. This mean is the arrival rate parameter for the Poisson distribution. By using the computed arrival rate and the corresponding cumulative Poisson distribution table we can determine the 80th percentile number of arrivals and therefore the required number of servers for that hour.

Your analysis should include the following activities:

4.1 Load the Bagel data file on your local computer system, and name the variables.

4.2 Compute the average number of arrivals for each hour, and convert that into an average number of arrivals for each three-minute service time.

4.3 Use a cumulative Poisson probability table or the routine on your computer to determine the number of arrivals just at or above the 80 percent CSL.

4.4 Using the required number of servers for each hour and the company goals for scheduling servers, work out a schedule that meets the CSL while using the minimum number of servers.

4.5 Prepare a written report that presents your scheduling plan and discusses its details.

NATIONAL COTTON FABRIC ASSOCIATION

Prediction of Quantity Demanded

5.1 INTRODUCTION AND BACKGROUND

George Castro, executive director of the National Cotton Fabric Association (NCFA), was asked to investigate the recent history of the industry. This action was prompted by a decline in production and sales of domestic cotton fabric. Some members of the board of directors claimed that the problem resulted from large increases in the volume of imported fabric on the market. Another group of board members suggested that price increases in recent years had encouraged clothing manufacturers to use other fabrics. The popularity of fabrics that require little or no ironing and involve easier care has also reduced sales of cotton fabric. George had observed both of these trends and believed that both were important. However, he also knew that he needed to present results from a rigorous study, in contrast to the general impressions he had gained from his casual observations. George asked the association's board of directors for authority to grant a contract to study the recent history and provide recommendations. The NCFA board approved George's request and named a subcommittee to meet with him to prepare a list of questions for the consultant to investigate.

A subcommittee from the board met with George to prepare specific questions in response to the concerns expressed at the board meeting. After considerable discussion, the group agreed that answers to the following questions should be obtained:

1. What has been the pattern of prices, production quantity, imported fabric quantity, and exported fabric quantity over the recent past?
2. What is the effect of price on the quantity of fabric produced?
3. Given the relationship of quantity versus price, what additional changes in production quantity are related to imports and to exports?
4. Given the NCFA's interest in maximizing total revenue, and working from recent experience, what price should be used?

TABLE 5.1	VARIABLE NAMES IN THE COTTON DATA FILE	

VARIABLE NAME	NUMBER OF OBSERVATIONS	VARIABLE DESCRIPTION
Year	28	Year for the observation
Quarter	28	Quarter for the observation
Cotprod	28	Total cotton production (in million lb)
Whoprice	28	Wholesale price for cotton (in 1967 $ per lb)
Import	28	Quantity of cotton fabric imported (in million lb)
Export	28	Quantity of cotton fabric exported (in million lb)

The subcommittee asked George to include these questions in the specifications for the study contract.

You have been awarded the contract to conduct the study for the NCFA. At an initial meeting, George reviewed the concerns and provided the list of questions prepared by the subcommittee. In addition, he gave you a data file named Cotton that contained important industry data.

After receiving the contract, you discussed the objectives with George and reviewed the questions. You concluded that the study should begin by establishing descriptive statistics for the variables, followed by developing a demand function (quantity versus price) using regression analysis. First, however, you must examine the data file description and identify the variables in Table 5.1.

The data file contains seasonally adjusted quarterly data obtained from industry sources. You also decide to review the economic background for demand functions. The results of that review are contained in the following section.

5.2 ECONOMIC BACKGROUND FOR DEMAND FUNCTIONS

Microeconomics develops the concept of demand curves and presents numerous applications for them. You have considered the theoretical derivation. It is possible to solve a number of economic problems qualitatively by using theoretically derived demand functions. In many applied business and economic problems, however, a specific mathematical relationship (equation) is desired. It is difficult to derive exact equations on the basis of economic theory alone. Therefore, demand functions are often derived empirically from historical data by using regression analysis. The purpose of this exercise is to introduce you to the process of empirical estimation.

METHODOLOGICAL NOTE

It is usually a bad idea to predict the quantity of demand by using independent variable values that lie outside the range of those given in the original data. The regression represents a "best fit" for the data used to derive the equation, but it does not necessarily fit the data outside of the range. Using the equation to predict values outside the range of the independent variables is called extrapolation. Extrapolation should only be done when you can safely assume that the relationship is the same outside the range of the independent variables as inside.

THEORETICAL PROBLEM

From microeconomic theory we know that a quantity sold—and hence purchased—is in equilibrium at the intersection of the supply and demand curves. If the curves did not shift, all points would have the same price and quantity, and it would be impossible to estimate a demand curve. Ideally, the estimation of a demand function would be performed under conditions where the demand curve (D) remained fixed and the supply curve (S) shifted as shown in Figure 5.1.

5.3   PROJECT REQUIREMENTS

After completing your analysis of the study goals and your review of the related background on demand functions, you believe that you are ready to proceed. Your final objective is to prepare a report to George Castro and the

FIGURE 5.1	DEMAND CURVE WITH SUPPLY CURVE SHIFTS

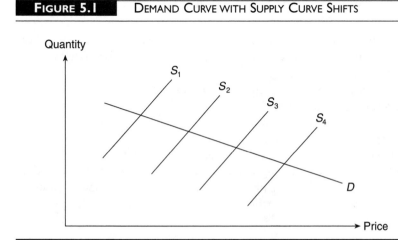

NCFA board that will respond to their concerns. They are particularly concerned about the effects of price and imports on cotton sales and production. In addition, they want you to use the demand models you develop to determine the price—within the range of observed prices—that will maximize total revenue in the industry. You must perform the following steps.

5.1 Load the data from the Cotton file into your local computer system, and name the variables.

5.2 Prepare time series plots of the four variables—cotton production, wholesale price, imported fabric, and exported fabric. Discuss the patterns of the variables over time.

5.3 Prepare a scatter plot for the observed values of cotton production and wholesale price, using your computer's statistical package. Discuss the relationship based on the observed plot pattern. Does the quantity demanded increase or decrease with increases in price?

5.4 Use simple least squares regression to estimate the relationship between quantity and price.

* **5.5** Plot the residuals from the least squares regression versus the predicted cotton production. Describe any unusual patterns in the data.

5.6 Determine the price that will yield the highest total revenue. [*Hint:* The regression equation computes quantity of cotton production as a function of price, and total revenue is price times quantity. Use your computer program to compute total revenue for each price, and plot total revenue versus price.]

* **5.7** Develop a mathematical function for total revenue as a function of price. Then use differential calculus to find the price that yields the "global maximum" total revenue. Explain why this price differs from the price you obtained in the previous question. Which price should you use in your report to George?

† **5.8** Estimate a model that predicts new production quantity as a function of imports and exports in addition to price. Use that function to answer the following questions.
   a. How much does the quantity produced increase for each unit of exported cotton fabric?
   b. How much does the quantity produced decrease for each unit of imported cotton fabric?
   c. What is the effect on cotton production when trade is balanced— that is when the quantity imported is equal to the quantity exported?
   d. Use the answers to the preceding questions to recommend a strategy for a trade policy that would benefit the cotton fabric industry

---

*These questions may require special material not covered in your course work to date.
†This question requires advanced understanding of multiple regression.

5.9 Prepare a two-page written report addressed to the NCFA Board, beginning with a short executive summary. This report should discuss your analysis and present the results requested for you study. You may attach additional graphs and tables to support your conclusions.

SPORTS UNLIMITED INC.

Analysis of Small Store Revenue

6.1 INTRODUCTION

George Jones is developing a chain of sporting goods stores in small towns across the Southeastern United States. Currently, he has a very successful set of five stores and hopes to expand by acquiring approximately twenty existing stores per year over the next five years. He has asked for your help to analyze the proposed purchase of Edwards Bros. Sports in Pine Hills, Alabama. George also requests that your analysis develop a general procedure that he can apply to other potential store purchases. He has obtained Edwards Bros. Sports' daily sales data for the past eighteen months. You are to use this data to study sales and revenue patterns.[1]

George Jones spent twenty years working for major regional retailers after receiving his MBA from a top-ten business school. His experience has included both financial and marketing executive positions. He also worked as a buyer and managed several stores. Two years ago a staff reduction occurred at the firm that employed him, as a result of which he received a generous severance payment equal to eighteen months of current salary. George took this lump-sum windfall as an opportunity to develop a business that would supply high-quality sporting goods at moderate prices to people in small towns.

During the past two years, George has acquired five stores and has developed a centralized purchasing operation that significantly reduces inventory cost. In addition he has developed a financial system and an inventory control system that maintain immediately accessible information on all items in stock. The inventory system supports routine analysis to determine the demand patterns for all items sold. As a result George has been able to select items that enjoy strong consumer demand, thereby increasing sales in all five stores.

---

[1]The data for this project were obtained from a study conducted by the author. All names and references are fictitious and should not be confused with any actual person or organization. The case does represent a typical real application.

Now that the initial operation with five stores is running smoothly, George is ready to begin the next phase of expansion by acquiring additional existing stores. At present, with five stores, George is involved in all decisions, including managing one of the stores. That experience has provided him with the opportunity to develop and test a number of successful operating procedures. Now he wants to implement these procedures in a management system and delegate responsibility. Included in this management decision system will be a procedure for analyzing potential store acquisitions.

6.2 ANALYSIS REQUIREMENTS

You have been asked to help design a procedure for analyzing store acquisitions. This procedure will be used to analyze daily sales revenue for the past one to two years. George believes that, with his management system in place, sales can be increased by at least 25 percent after two years. However, he does not wish to acquire stores that are not at least breaking even. In addition he wants to know how sales vary over the year, so that cash flow and long-term staffing issues can be resolved.

Your initial task is to develop an analysis procedure, using the historical daily sales revenue. George has asked that you analyze the data from the sporting goods store in Pine Hills, Alabama, and use that analysis as a model for analyzing other proposed acquisitions.

Discussion with George provides a basic understanding of the operating costs for sporting goods stores. Costs of store operations can be divided into fixed and variable costs. Fixed costs include items such as rent, insurance, interest cost on store fixtures and inventory stock, manager's salary and benefits, data-processing and accounting charges, and community development. Variable costs include extra sales staff and advertising, among other things. You should consult a text on retail accounting for additional details. The financial analysis model includes a cost component that has the following form:

Total cost = Cost of goods sold + Total fixed costs + Total variable costs

$$C_T = C_g + C_f + C_v$$

The model also includes a total revenue component derived from goods sold:

R = Total revenue = Total receipts from all goods sold

Store profit is computed as total revenue minus total cost.

Daily sales revenue data for the past eighteen months are stored in a data file named Sports. The variables are described in Table 6.1.

TABLE 6.1	VARIABLE NAMES IN THE SPORTS DATA FILE	
VARIABLE NAME	**NUMBER OF OBSERVATIONS**	**VARIABLE DESCRIPTION**
Year	468	Year in which the sales occurred
Month	468	Month of sales (1 = January, 12 = December)
Dayweek	468	Day of week (1 = Monday, 6 = Saturday)
Salesrev	468	Daily sales revenue (in dollars)

The first step in your analysis is to determine the expected annual sales revenue and the amount available to cover fixed and variable expenses for store operations and to provide a contribution to profit. The store will be open 300 days per year—an average of 25 days per month. Using the estimated mean and estimated variance for daily sales revenue, you can compute the mean and the variance for annual and monthly sales revenue. Using these means and variances you can calculate a 95 percent lower confidence interval for the estimates.

Another important issue in the financial management of stores is the variation in cash flow over the year. Each month a number of costs must be covered from the revenue obtained during that month. If sales revenues are not sufficient, then a shot-term loan must be obtained to cover the monthly costs not paid from sales revenue. Thus George wants to know the expected pattern of sales over the various months. Using this he can determine the number of months during which short-term loans will be needed, as well as the anticipated size of each loan.

After discussion with George, you agree that your study should include the following steps.

6.1 Determine the annual and average monthly contributions to profit and store operating costs, based on the present daily sales revenue.

6.2 Prepare an analysis of monthly cash flow over the year that can be used to cover store operations for the proposed store.

* **6.3** Develop a model to forecast daily sales with adjustments for day of week and month. [*Note:* This model can be developed by using dummy variables and multiple regression.]

6.4 Prepare a written report that discusses your analysis of the proposed acquisition and the values of key measures for the Pine Hills store. A second part of the report should describe the analysis procedure and key measures that would be used to analyze all potential store acquisitions.

*This question requires advanced understanding of multiple regression.

CONSOLIDATED FOODS INC. A

Market Description

7.1 RETAIL MARKET DESCRIPTION

Susan Thurston, product manager for Consolidated Foods Inc., is responsible for developing and evaluating promotional campaigns to increase sales for her canned food product. She has just been appointed product manager for Brand 1, which competes in retail food markets with four other major brands. The product is well established in a stable market and is generally viewed by consumers as a food commodity. Successful product performance requires strategies that encourage customers to move to Susan's brand from the various competing brands. She can encourage this movement by various combinations of reduced price specials, in-store displays, and newspaper advertising (partial support for a weekly supermarket ad that emphasizes Brand 1). Her competitors have at their disposal the same strategic tools for increasing their sales. By successfully manipulating these marketing variables relative to the actions of her competitors, Susan works to develop a successful sales performance record, as demonstrated by a stable or increasing market share.

Susan believes that her future success depends on her knowledge of the market. Therefore, she has asked for your help in analyzing the past year of weekly data. Your task is to provide Susan with information and a solid understanding of the market. You will apply basic statistical procedures to the data and obtain important descriptive statistics. Then you will perform your analysis of the data and provide a written description of the conclusions from your work. From your report, Susan will gain an understanding of the market so that she can develop optimal marketing strategies.

---

[1]The data for this study were collected from supermarket sales records by a Nielsen Marketing Research project for a food product. These data, from a Nielsen SCANTRACK Major Market, were generously supplied to the author for use in teaching statistics; however, we are not allowed to reveal the name of the product or the supermarkets because the data remain proprietary. Nielsen SCAN-TRACK Major Markets are the geographic regions or markets in which Nielsen SCANTRACK U.S. is established. This service provides weekly scanning-based reports on sales and prices in large ($4,000,000 and greater annual volume) supermarkets. The data are real, and the analyses conducted in this case are typical marketing analysis applications.

| **TABLE 7.1** | VARIABLE LIST FOR WEEKLY STORE SALES DATA IN THE CONFOOD DATA FILE | | |

VARIABLE NUMBER	VARIABLE NAME	COUNT	VARIABLE DESCRIPTION
1	Storenum	156	Code number for supermarket (coded 1, 2, or 3)
2	Weeknum	156	Consecutive week number
3	Saleb1	156	Total unit sales for Brand 1
4	Apriceb1	156	Actual retail price for Brand 1
5	Rpriceb1	156	Regular or recommended price for Brand 1
6	Promotb1	156	Promotion code for Brand 1
			0 = No promotion
			1 = Newspaper advertising only
			2 = In-store display only
			3 = Newspaper ad and in-store display
7	Saleb2	156	Total unit sales for Brand 2
8	Apriceb2	156	Actual retail price for Brand 2
9	Rpriceb2	156	Regular or recommended price for Brand 2
10	Promotb2	156	Promotion code for Brand 2
11	Saleb3	156	Total unit sales for Brand 3
12	Apriceb3	156	Actual retail price for Brand 3
13	Rpriceb3	156	Regular or recommended price for Brand 3
14	Promotb3	156	Promotion code for Brand 3
15	Saleb4	156	Total unit sales for Brand 4
16	Apriceb4	156	Actual retail price for Brand 4
17	Rpriceb4	156	Regular or recommended price for Brand 4
18	Promotb4	156	Promotion code for Brand 4
19	Saleb5	156	Total unit sales for Brand 5
20	Apriceb5	156	Actual retail price for Brand 5
21	Rpriceb5	156	Regular or recommended price for Brand 5
22	Promotb5	156	Promotion code for Brand 5

Supermarket sales data for analysis and evaluation are routinely purchased from a national marketing research company.[1] Your analysis will use data from three major supermarket chains that together distribute 60 percent of the product. The marketing research department has obtained weekly brand tracking data for the three chains for the past year. The variable list is shown in Table 7.1, and the data are stored in a file named Confood. These data include weekly sales, price, and promotion data for Susan's product and its four major competitors. It is collected from weekly sales audits in specific supermarkets, whose sales patterns follow national trends. Data from supermarkets in the same distribution chain are aggregated.

A series of case project analyses based on these data will be conducted. The analyses will be assigned as the necessary analytical skills are developed. A typical marketing analysis would logically include most of the different steps contained in the set of all four Consolidated Foods case projects presented in this casebook.

7.2   UNIVARIATE DATA ANALYSIS

Susan has first asked you to help her determine how to make the best use of the supermarket sales audit data that are regularly purchased. The spectrum of possible analyses ranges from a simple description of the variables to development of regression models that predict sales as a function of price and promotion variables.

The first step in any analysis is to obtain a description of the system represented by the data. This is usually done by computing descriptive statistics and preparing graphs that indicate the central tendency, dispersion, and pattern of sample data for each variable.

In addition, in many cases business and economic data are collected over time. Therefore, description may include graphical displays and time sequence analysis to show the variation of data over time. Based on this discussion, Susan requests that you prepare a descriptive analysis of the sales audit data and report your conclusions.

You begin by preparing a descriptive analysis of weekly sales for her product (Brand 1) and for the competing products (Brand 2 through Brand 5). This analysis will give Susan a baseline description and an indication of typical ranges for the key market performance and strategy variables.

The first step is to prepare univariate descriptions of key variables. These descriptions usually include a measure of central tendency (mean or median) and a measure of dispersion (standard deviation, variance, range, or quartiles). In addition, graphical displays such as histogram bar charts or box and whisker plots are often prepared. Susan is particularly interested in knowing about the following variables: Brand 1 sales (number of units sold) and market share (percentage of sales for each brand), sales of competing products and their market share, total market sales, Brand 1 actual price and total revenue, actual price and total revenue for each of the other products, and weighted average price for the other four brands.

The second step of the analysis is to examine the preceding measures over time. Because markets are dynamic, a sophisticated analysis requires examination of variations over time for the key market variables. Whenever we perform analyses over time, we need to carefully choose a length of time for each observation. A short time interval—a week, for these data—will provide considerable detail about the variation over time. However, it may also reveal fluctuations due to effects that cannot be measured or controlled and that follow a random pattern. In that case, examination of graphical time series plots can lead to considerable confusion. Therefore, analysts sometimes choose a longer time interval and compute a mean for each time interval observed. This strategy reduces random effects, but it may also hide important changes.

For this project you will plot the variables being studied against both the week number variable and a variable that groups each consecutive four-week period. In that second comparison, the data will be separated into thirteen four-week periods, and the means for each period will be computed and displayed graphically. You will compare total sales patterns with price patterns for Brand 1 and the competing brands, and discuss these in your report. You will have to decide which interval or combination of intervals provides the best analysis tool.

The specific requirements for this case project are as follows.

7.1 Use your local computer system to load the data file named Confood. The variables are described in Table 7.1. Then prepare appropriate data transformations. Your teacher will provide specific instructions for accessing the data file from your local computer system.

7.2 Compute appropriate descriptive statistics for the variables that describe the market. These variables include sales for each brand, total sales for the market, actual price for each brand, total sales for Brands 2 through 5, weighted average price for Brands 2 through 5, total sales revenue for each brand, and market share for each brand. Prepare histograms for total Brand 1 sales and for Brands 2 though 5 sales combined.

7.3 Prepare a time series analysis for Brand 1 sales, Brand 1 price, total sales for Brands 2 through 5, and weighted average price for Brands 2 through 5. Do this first by using the weekly data. Then group the fifty-two weeks into thirteen four-week periods. Your statistical computer package has a procedure for grouping observations and for computing means of grouped observations. Arrange to have the group mean computed and displayed for each consecutive time period.

7.4 Prepare a written report that describes the Brand 1 market for Susan. Draw comparisons between Brand 1 and its competitors. Refer in the report to the computed summary statistics and graphs you prepared earlier. Your report should emphasize key patterns and trends, and not merely discuss the detailed results, which should be included as appendices.

CONSOLIDATED FOODS INC. B

Price and Quantity Sold Relationships

8.1 RETAIL FOOD SALES ANALYSIS

Susan Thurston, product manager for Consolidated Foods Inc., is responsible for developing and evaluating promotional campaigns to increase sales for her canned food product. She has just been appointed product manager for Brand 1, which competes in retail food markets with four other major brands. The product is well established in a stable market and is generally viewed by consumers as a food commodity. Successful product performance requires strategies that encourage customers to move to Susan's brand from the various competing brands. She can encourage this movement by various combinations of reduced price specials, in-store displays, and newspaper advertising (partial support for a weekly supermarket ad that emphasizes Brand 1). Her competitors have at their disposal the same strategic tools for increasing their sales. By successfully manipulating these marketing variables relative to the actions of her competitors, Susan works to develop a successful sales performance record, as demonstrated by a stable or increasing market share.

Susan believes that her future success depends on her knowledge of the market. Therefore, she has asked for your help in analyzing the past year of weekly data. Your task is to provide Susan with information and a solid understanding of the market. You will apply basic statistical procedures to the data and obtain descriptions of simple relationships between the number of units sold and the price charged per unit. Then you will perform your analysis of the data and provide a written description of the conclusions from your work. From your report, Susan will gain an understanding of the market so that she can develop optimal marketing strategies.

Supermarket sales data for analysis and evaluation are routinely purchased from a national marketing research company.[1] Your analysis will use data

---

[1]The data for this study were collected from supermarket sales records by a Nielsen Marketing Research project for a food product. These data, from a Nielsen SCANTRACK Major Market, were generously supplied to the author for use in teaching statistics; however, we are not allowed to reveal the name of the product or the supermarkets because the data remain proprietary. Nielsen SCAN-

TABLE 8.1	VARIABLE LIST FOR WEEKLY STORE SALES DATA IN THE CONFOOD DATA FILE

VARIABLE NUMBER	**VARIABLE NAME**	**COUNT**	**VARIABLE DESCRIPTION**
1	Storenum	156	Code number for supermarket (coded 1, 2, or 3)
2	Weeknum	156	Consecutive week number
3	Saleb1	156	Total unit sales for Brand 1
4	Apriceb1	156	Actual retail price for Brand 1
5	Rpriceb1	156	Regular or recommended price for Brand 1
6	Promotb1	156	Promotion code for Brand 1
			0 = No promotion
			1 = Newspaper advertising only
			2 = In-store display only
			3 = Newspaper ad and in-store display
7	Saleb2	156	Total unit sales for Brand 2
8	Apriceb2	156	Actual retail price for Brand 2
9	Rpriceb2	156	Regular or recommended price for Brand 2
10	Promotb2	156	Promotion code for Brand 2
11	Saleb3	156	Total unit sales for Brand 3
12	Apriceb3	156	Actual retail price for Brand 3
13	Rpriceb3	156	Regular or recommended price for Brand 3
14	Promotb3	156	Promotion code for Brand 3
15	Saleb4	156	Total unit sales for Brand 4
16	Apriceb4	156	Actual retail price for Brand 4
17	Rpriceb4	156	Regular or recommended price for Brand 4
18	Promotb4	156	Promotion code for Brand 4
19	Saleb5	156	Total unit sales for Brand 5
20	Apriceb5	156	Actual retail price for Brand 5
21	Rpriceb5	156	Regular or recommended price for Brand 5
22	Promotb5	156	Promotion code for Brand 5

from three major supermarket chains that together distribute 60 percent of the product. The marketing research department has obtained weekly brand tracking data for the three chains for the past year. The variable list is shown in Table 8.1, and the data are stored in a file named Confood. These data include weekly sales, price, and promotion data for Susan's product and its four major competitors. It is collected from weekly sales audits in specific supermarkets, whose sales patterns follow national trends. Data from supermarkets in the same distribution chain are aggregated.

A series of case project analyses based on these data will be conducted. The analyses will be assigned as the necessary analytical skills are developed. A typical marketing analysis would logically include most of the different steps contained in the set of all four Consolidated Foods case projects presented in this casebook.

TRACK Major Markets are the geographic regions or markets in which Nielsen SCAN-TRACK U.S. is established. This service provides weekly scanning-based reports on sales and prices in large ($4,000,000 and greater annual volume) supermarkets. The data are real, and the analyses conducted in this case are typical marketing analysis applications.

8.2 DESCRIPTIVE RELATIONSHIPS: PRICE, PROMOTION, AND QUANTITY

Univariate descriptive relationships, such as those prepared in Consolidated Foods Inc. A (Case 7) indicate the levels and ranges of important market variables. Thus they identify typical conditions and ranges for normal operations of the market. They do not, however, reveal how one variable changes in response to changes in other variables. For example, does the quantity of Brand 1 sold change in response to changes in the price of Brand 1? Does it change in response to changes in the price of other brands? Answers to these questions provide a basis for developing marketing strategies designed to increase total sales.

After reviewing your descriptive analysis, Susan asks about relationships between sales and the various price and promotion variables. How much do quantities sold change when the actual retail price is lowered? Do the quantities sold increase after newspaper advertising? Do total sales for all brands change when the average price for all brands changes? Do total sales for Brand 1 change with changes in the average price for all competing brands? These questions deal with simple relationships between variables—relationships that can be described by graphical plots, sample correlation coefficients, and simple least squares regression lines. These descriptive measures have considerable intuitive appeal and in many cases supply important insights into the system represented by the data.

Relationships between business and economic variables often involve high-order simultaneous relationships between more than two variables. In those case, graphical plots and correlation coefficients may provide confusing and even misleading results. For example, the simple relationship between price and quantity sold for a product may be clouded by simultaneous changes in the price of competing products. Thus one does not know whether consumers are responding to the price change for the product of interest or to price changes for competing products. These simultaneous price effects may cancel each other.

8.3 THEORY NOTE

An example of confusing relationships between variables involves the relationship between quantity supplied and price and the relationship between quantity demanded and price in a typical market. Economic theory tells us that the combination of price and quantity sold results from the intersection of supply and demand functions. Moreover, our understanding of statistics suggests that the actual observed quantity sold should equal the intersection or equilibrium sales plus or minus a random error term. An ideal demand

| **FIGURE 8.1** | QUANTITY VS. PRICE, AND DEMAND VS. SHIFTING SUPPLY |

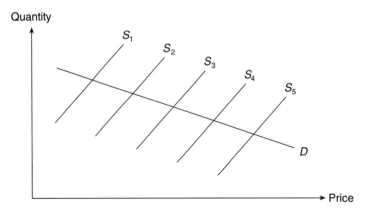

function is represented by the data in Figure 8.1. For this hypothetical example, the data indicate a generally decreasing level of sales with higher prices. From economic theory we know that this pattern would result from a stable demand relationship with a shifting supply function. Thus the data result from the equilibrium between a single demand function and a set of shifted supply functions. You will use this assumption for your analysis.

However, other models of the economic process could be used. For example, the data could have resulted from a stable supply function and a shifting demand function. Both supply and demand functions might shift at the same time. In that case the observed equilibrium points do not correspond to either a supply function or a demand function. When this condition occurs, additional variables are needed to estimate the simultaneous shifts in the supply and demand relationships. In Case 10 we will make such an adjustment, but only after you have studied multiple regression.

8.4 SPECIFIC REQUIREMENTS

After discussing matters with Susan, you agree to undertake the following steps.

8.1 Load the Confood data file, which is described in Table 8.1. Use the transformation commands in your statistical package to compute the weighted average price for Brands 2 through 5 (the brands that compete with Brand 1) for each time period. Your professor will provide the instructions required for your computer system.

8.2 Compute correlation coefficients to describe the linear relationship between:

    a. Sales—number of units sold—and price for each of the five brands.

    b. Brand 1 sales and weighted average price for Brands 2 through 5.

    c. All combinations of prices for the five brands.

    d. All combinations of quantity sold for the five brands.

8.3 Prepare scatter plots for sales versus price for all five brands.

8.4 Prepare a scatter plot for Brand 1 sales versus the weighted average price of all competing brands.

8.5 Use simple regression analysis to describe the linear relationships for Brand 1 sales versus Brand 1 price and for Brand 1 sales versus the weighted average prices of all competing brands.[2]

8.6 Prepare an analysis to indicate the relationship between quantity sold and type of promotion used.

8.7 Prepare a written discussion of the relationship between quantity sold and prices, based on the results of your analysis. This report should include recommendatons for developing marketing strategies.

---

[2]You are only required to obtain the constant and slope coefficient for the linear relationship between price and quantity sold, using the simple regression option on your local computer package. At this point you are seeking a description of the function and are not required to analyze the remainder of the regression output.

CONSOLIDATED FOODS INC. C

Comparison of Market Sales

9.1   RETAIL FOOD SALES ANALYSIS

Susan Thurston, product manager for Consolidated Foods Inc., is responsible for developing and evaluating promotional campaigns to increase sales for her canned food product. She has just been appointed product manager for Brand 1, which competes in retail food markets with four other major brands. The product is well established in a stable market and is generally viewed by consumers as a food commodity. Successful product performance requires strategies that encourage customers to move to Susan's brand from the various competing brands. She can encourage this movement by various combinations of reduced price specials, in-store displays, and newspaper advertising (partial support for a weekly supermarket ad that emphasizes Brand 1). Her competitors have at their disposal the same strategic tools for increasing their sales. By successfully manipulating these marketing variables relative to the actions of her competitors, Susan works to develop a successful sales performance record, as demonstrated by a stable or increasing market share.

Susan believes that her future success depends on her knowledge of the market. Therefore, she has asked for your help in analyzing the past year of weekly data. Your task is to provide Susan with information and a solid understanding of the market. You will compare the sales performance of Brand 1 with that of its market competitors. Then you will develop your analysis and provide a carefully written description of the conclusions you have drawn from your work. From your report, Susan will gain an understanding of the market so that she can develop optimal marketing strategies.

---

[1]The data for this study were collected from supermarket sales records by a Nielsen Marketing Research project for a food product. These data, from a Nielsen SCANTRACK Major Market, were generously supplied to the author for use in teaching statistics; however, we are not allowed to reveal the name of the product or the supermarkets because the data remain proprietary. Nielsen SCAN-TRACK Major Markets are the geographic regions or markets in which Nielsen SCANTRACK U.S. is established. This service provides weekly scanning-based reports on sales and prices in large ($4,000,000 and greater annual volume) supermarkets. The data are real, and the analyses conducted in this case are typical marketing analysis applications.

Supermarket sales data for analysis and evaluation are routinely purchased from a national marketing research company.[1] Your analysis will use data from three major supermarket chains that together distribute 60 percent of the product. The marketing research department has obtained weekly brand tracking data for the three chains for the past year. The variable list is shown in Table 9.1, and the data are stored in a file named Confood. These data include weekly sales, price, and promotion data for Susan's product and its four major competitors. It is collected from weekly sales audits in specific supermarkets, whose sales patterns follow national trends. Data from supermarkets in the same distribution chain are aggregated.

A series of case project analyses based on these data will be conducted. The analyses will be assigned as the necessary analytical skills are developed. A typical marketing analysis would logically include most of the different steps contained in the set of all four Consolidated Foods case projects presented in this casebook.

9.2 COMPARING SALES FOR COMPETING BRANDS

A report on last year's sales performance is being prepared, and you have been asked to assist with the statistical analysis. Nina Gupta, Confood's vice president of marketing, has asked for a report on the sales status of Brand 1 in comparison with the four competing brands. This report will be used to determine product sales performance relative to the rest of the market. She has asked for an estimate of average weekly sales (number of units sold), average weekly price, market share (percentage of units sold by each brand), and average weekly sales revenue. In addition, she has asked for a comparison of Brand 1 with Brands 2 and 3 with respect to these measures.

After discussion with Susan you conclude that a combination of classical estimation and hypothesis testing will provide the desired statistical results for your study. Therefore, you agree that you will first prepare 95 percent confidence intervals for the four measures requested by Ms. Gupta.

In the past, Brand 2 has been the leading seller, and Brand 1 has usually been ahead of Brand 4 in total sales and revenue. You decide to perform a test to determine whether Brand 2 is still ahead of Brand 1 in average weekly sales and average weekly revenue. A second test will be used to determine whether Brand 1 remains ahead of Brand 4 on these measures. You agree to design and implement appropriate hypothesis tests to answer these questions. This task involves the following steps.

9.1 Load the Confood data file which is described in Table 9.1

9.2 Determine the average and the variation for Brand 1 price, sales, market share, and total revenue, using the weekly market survey data.

TABLE 9.1			VARIABLE LIST FOR WEEKLY STORE SALES DATA IN THE CONFOOD DATA FILE
VARIABLE NUMBER	**VARIABLE NAME**	**COUNT**	**VARIABLE DESCRIPTION**
1	Storenum	156	Code number for supermarket (coded 1, 2, or 3)
2	Weeknum	156	Consecutive week number
3	Saleb1	156	Total unit sales for Brand 1
4	Apriceb1	156	Actual retail price for Brand 1
5	Rpriceb1	156	Regular or recommended price for Brand 1
6	Promotb1	156	Promotion code for Brand 1
			0 = No promotion
			1 = Newspaper advertising only
			2 = In-store display only
			3 = Newspaper ad and in-store display
7	Saleb2	156	Total unit sales for Brand 2
8	Apriceb2	156	Actual retail price for Brand 2
9	Rpriceb2	156	Regular or recommended price for Brand 2
10	Promotb2	156	Promotion code for Brand 2
11	Saleb3	156	Total unit sales for Brand 3
12	Apriceb3	156	Actual retail price for Brand 3
13	Rpriceb3	156	Regular or recommended price for Brand 3
14	Promotb3	156	Promotion code for Brand 3
15	Saleb4	156	Total unit sales for Brand 4
16	Apriceb4	156	Actual retail price for Brand 4
17	Rpriceb4	156	Regular or recommended price for Brand 4
18	Promotb4	156	Promotion code for Brand 4
19	Saleb5	156	Total unit sales for Brand 5
20	Apriceb5	156	Actual retail price for Brand 5
21	Rpriceb5	156	Regular or recommended price for Brand 5
22	Promotb5	156	Promotion code for Brand 5

9.3 Perform a statistical test to determine whether Brand 2 continues to have higher sales than Brand 1. The test should be designed to provide strong evidence that Brand 2 sales are higher if they are.

9.4 Perform a statistical test to determine whether Brand 1 is still ahead of brand 4 average weekly sales.

9.5 Write a one-page report to Ms. Gupta presenting your conclusions regarding the market performance of Brand 1. This report should clearly indicate the sales performance for Brand 1 and should compare it to the performance of the competition. In an appendix, you may include tables and graphs to show the results of your analysis and help communicate your conclusions.

CONSOLIDATED FOODS INC. D

Prediction of Sales

10.1 RETAIL FOOD SALES ANALYSIS

Susan Thurston, product manager for Consolidated Foods Inc., is responsible for developing and evaluating promotional campaigns to increase sales for her canned food product. She has just been appointed product manager for Brand 1, which competes in retail food markets with four other major brands. The product is well established in a stable market and is generally viewed by consumers as a food commodity. Successful product performance requires strategies that encourage customers to move to Susan's brand from the various competing brands. She can encourage this movement by various combinations of reduced price specials, in-store displays, and newspaper advertising (partial support for a weekly supermarket ad that emphasizes Brand 1). Her competitors have at their disposal the same strategic tools for increasing their sales. By successfully manipulating these marketing variables relative to the actions of her competitors, Susan works to develop a successful sales performance record, as demonstrated by a stable or increasing market share.

Susan believes that her future success depends on her knowledge of the market. Therefore, she has asked for your help in analyzing the past year of weekly data. Your task is to provide Susan with information and a solid understanding of the market. Specifically, you are to determine the influence of various factors on the sale of Brand 1. Then you are to perform your analysis of the data and provide a written description of the conclusions from your work. From your report, Susan will gain an understanding of the market so that she can develop optimal marketing strategies.

---

[1]The data for this study were collected from supermarket sales records by a Nielsen Marketing Research project for a food product. These data, from a Nielsen SCANTRACK Major Market, were generously supplied to the author for use in teaching statistics; however, we are not allowed to reveal the name of the product or the supermarkets because the data remain proprietary. Nielsen SCANTRACK Major Markets are the geographic regions or markets in which Nielsen SCANTRACK U.S. is established. This service provides weekly scanning-based reports on sales and prices in large ($4,000,000 and greater annual volume) supermarkets. The data are real, and the analyses conducted in this case are typical marketing analysis applications.

Supermarket sales data for analysis and evaluation are routinely purchased from a national marketing research company.[1] Your analysis will use data from three major supermarket chains that together distribute 60 percent of the product. The marketing research department has obtained weekly brand tracking data for the three chains for the past year. The variable list is shown in Table 10.1, and the data are stored in a file named Confood. These data include weekly sales, price, and promotion data for Susan's product and its four major competitors. It is collected from weekly sales audits in specific supermarkets, whose sales patterns follow national trends. Data from supermarkets in the same distribution chain are aggregated.

A series of case project analyses based on these data will be conducted. The analyses will be assigned as the necessary analytical skills are developed. A typical marketing analysis would logically include most of the different steps contained in the set of all four Consolidated Foods case projects presented in this casebook.

10.2 PREDICTION OF QUANTITY DEMANDED

Susan is developing her Brand 1 promotional plan for the coming year. She has asked you to prepare a study of the sales effect of the various promotional variables over the past year. Promotions to increase sales consist primarily of price reductions, in-store displays, and newspaper advertising.

The effect of price is expected to follow basic economic theory concerning supply and demand. This theory concludes that, for most products, the quantity demanded is inversely related to the product price and is directly related to the price of competing products. Each brand in the market has a regular, established price. During various weeks, price specials are used to stimulate sales. If Brand 1 is the only product with a reduced price, consumers are expected to switch their purchase from other brands and increase purchases of Brand 1. In some cases brand loyalty will overcome short-term specials, and customers will not switch. If customers view the product as a commodity (such that each brand has the same quality) customers will usually purchase the brand with the lowest price. Thus we expect that the quantity of Brand 1 sold during a particular week will be inversely related to the price of Brand 1 and directly related to the price of competing brands.

The effect of in-store displays and newspaper advertising is more complicated to assess. Newspaper advertising for a product usually involves partial support for a store's weekly advertising, with the understanding that the product will be featured in the advertisement. In many cases advertising copy is supplied to the store for use in the ad. This ad may be combined with a price special. Store owners are also encouraged—usually with financial incentives—to provide special displays and/or locations that feature the advertised brand in contrast to other brands. The rationale for spending

TABLE 10.1	VARIABLE LIST FOR WEEKLY STORE SALES DATA IN THE CONFOOD DATA FILE		

VARIABLE NUMBER	VARIABLE NAME	COUNT	VARIABLE DESCRIPTION
1	Storenum	156	Code number for supermarket (coded 1, 2, or 3)
2	Weeknum	156	Consecutive week number
3	Saleb1	156	Total unit sales for Brand 1
4	Apriceb1	156	Actual retail price for Brand 1
5	Rpriceb1	156	Regular or recommended price for Brand 1
6	Promotb1	156	Promotion code for Brand 1
			0 = No promotion
			1 = Newspaper advertising only
			2 = In-store display only
			3 = Newspaper ad and in-store display
7	Saleb2	156	Total unit sales for Brand 2
8	Apriceb2	156	Actual retail price for Brand 2
9	Rpriceb2	156	Regular or recommended price for Brand 2
10	Promotb2	156	Promotion code for Brand 2
11	Saleb3	156	Total unit sales for Brand 3
12	Apriceb3	156	Actual retail price for Brand 3
13	Rpriceb3	156	Regular or recommended price for Brand 3
14	Promotb3	156	Promotion code for Brand 3
15	Saleb4	156	Total unit sales for Brand 4
16	Apriceb4	156	Actual retail price for Brand 4
17	Rpriceb4	156	Regular or recommended price for Brand 4
18	Promotb4	156	Promotion code for Brand 4
19	Saleb5	156	Total unit sales for Brand 5
20	Apriceb5	156	Actual retail price for Brand 5
21	Rpriceb5	156	Regular or recommended price for Brand 5
22	Promotb5	156	Promotion code for Brand 5

money on these promotions is that supplying information about the product increases the chance of purchase. However, information from other advertisements, personal communications, subjective reaction to previous use of the brand, and other factors can strengthen or weaken the effect of immediate advertising and promotion.

The plan that Susan is developing will indicate how she can allocate her marketing budget. Price specials are "purchased" from the supermarket manager by reducing the price charged for the product, with the understanding that the retail price will be lowered. Alternatively, in-store promotions and/or newspaper advertising is purchased by a payment to the store that reduces the total cost of purchasing a shipment of (in this case) Brand 1. In many cases, combinations of the variables are used. For example, a newspaper ad may be used to publicize a price special for Brand 1. This combination may be cheaper than the sum of the price special and newspaper advertising, because the supermarket manager is using the special in the ad to draw customers to the store.

In this project you will use regression analysis to determine the effect of price and other promotional strategy variables on sales performance. The regression models will predict both weekly sales and market share for Brand 1 in the stores sampled. Thus, the results will indicate both the effect on increasing sales and the effect on drawing sales away from competing products. Included in the list of independent predictor variables should be the price for Brand 1 and the weighted average price for all competing brands. The various promotions can be included by using dummy variables. If weekly sales differ among the three store chains, dummy variables should be used to adjust the mean sales in the regression model that predicts total weekly sales. Review the procedures for computing new variables in your statistical computer package. Then perform the following steps.

10.1 Load the Confood data file, which is described in Table 10.1.

10.2 Compute the independent and dependent variables needed for your regression models, using the transformation routines in your statistical package. Convert the promotion variable to an appropriate set of dummy variables.

10.3 Develop a model to predict weekly sales and market share for Brand 1. Select only statistically significant variables for your final model, by using appropriate analysis procedures.

10.4 Examine the deviations of individual weeks from the model, and attempt to identify special circumstances that led to higher or lower sales than expected.

10.5 Prepare a written report—including the final model in standard form—that discusses the effect of the promotional and price variables on sales. Based on your analysis, make recommendations and draw conclusions that will help Susan develop her plan. Since you do not know the cost of each promotion, you cannot recommend specific options. However, you should be able to indicate the expected outcome if Susan uses a particular option. Include numerical examples based on the regression models.

MIDWEST HEAVY EQUIPMENT INC. A

Description of Receivables

11.1 INTRODUCTION

Midwest Heavy Equipment Inc. is a major regional distributor of large construction equipment. The company was started by Oscar Thompson in the 1950s to service contractors who were building the interstate highway system. From the beginning, Midwest sold a wide range of equipment to contractors throughout the region. The company maintained a large stock of new and used equipment and established a reputation for reliably supplying contractors with their equipment needs. A large-scale service operation provided both on-site and shop-based repair services. In the beginning, many of the contractors were starting or expanding rapidly as they obtained a number of lucrative contracts.

Oscar Thompson established a reputation of support for new contractors, offering them quick access to equipment and services. As a result, these new companies did not need to buy various extra pieces of equipment that then sat idle while waiting for repair or for assignment to some incoming contract. Oscar also extended credit generously and did not pressure new firms for rapid payment. In the early days, he negotiated low-cost lines of credit with a number of large banks to cover the extended credit to his customers.

Because of its high service level to growing firms, Midwest did not need to compete on price. Midwest grew to be one of the largest suppliers of construction equipment in the region. The company employs over 3,000 people in its central facility in Minneapolis, and in fifty service and supply facilities from Ohio to Colorado and from Minnesota to Texas. Most contractors remember the high-quality service and attractive payment extensions provided by Oscar Thompson in the early years of their operation. Therefore, Midwest Equipment has had a large following of loyal customers. Midwest was a very profitable business through the 1960s and most of the 1970s. Oscar Thompson became a wealthy man and contributed generously to a number of colleges and other needs in the region.

In the mid-1970s, Oscar's daughter, Megan, completed her MBA at a top-ten school and began working with the company. Oscar had always planned that she would take over the company when he retired. During her college years she worked at a number of jobs, including mechanics helper, customer service representative, and financial operations associate. She often accompanied her father when he called on major customers, and she also drove a truck to deliver equipment. After obtaining her MBA, Megan worked in a number of marketing, financial, and management positions.

During the late 1970s and early 1980s, Midwest Heavy Equipment began to feel pressures similar to those experienced by a number of businesses. Foreign manufacturers of heavy equipment were setting up their own distribution facilities in direct competition with Midwest. These companies sold high-quality equipment at prices generally below those charged by Midwest. Contractors tended to be better established, and the competition between contractors was severe. Therefore, contractors sought to reduce costs any way they could. In addition, interest rates rose to high levels. This greatly increased the debt service cost of Midwest's line of credit, which it had traditionally used to cover inventory and accounts receivable.

11.2 DESCRIPTION OF RECEIVABLES

Megan Thompson, now Midwest's vice president of finance and accounting, wants to know the size and distribution of accounts receivable. That knowledge will help her determine both the potential benefits from a well-managed receivables policy and the savings achievable through a reduction in the total dollar value of receivables.

As a first step she has asked you to conduct a statistical study of accounts receivable. A random sample of receivables will be used to describe the age and dollar value of al receivables. The receivable records are contained in the data file named Midwest.[1]

The variables to be used in this study are as follows:

Ident: Receivable record identification number.

Pstduday: Number of days the receivable is past its due date.

Bookval: Dollar value entered into the receivable record file.

Auditval: Actual dollar value as determined by the external auditors.

After discussing the situation with Megan, you agree to perform the following steps.

---

[1]The data for this study were generously supplied by Ernst & Young. Sincere thanks to Mark Hornung, senior manager of the Minneapolis office, and to Mike Giese from the company's national office in Cleveland.

11.1 Load the data from the Midwest file into your computer system.

11.2 Prepare descriptive statistics and graphs for the number of days the receivables are past due.

    a. Describe the distribution and indicate the quartile points.

    b. Is the distribution skewed or symmetric?

    c. What is the mean number of days past due?

11.3 Prepare descriptive statistics and a graph for the book value and the audit value of the receivables.

    a. Briefly describe the distribution of the value of the receivables.

    b. What is the mean value of the receivables?

11.4 Using your computer package, compute a new variable that represents the difference between the book value and the audit value of each receivable.

    a. Describe the distribution of the difference between book value and audit value.

    b. What percentage of receivables have a difference between the book value and the audit value? [*Hint:* A receivable with a difference has a nonzero value for the difference variable, and a receivable with no difference has a zero for the difference variable. You need to find an efficient way to count either the zeros or the nonzeros.]

11.5 Write a one-page report for Megan presenting the key descriptions of the sample of receivables. Emphasize the results that are important for her understanding of the receivables problem and for their eventual reduction.

MIDWEST HEAVY EQUIPMENT INC. B

Detecting Errors in Receivables

12.1 INTRODUCTION

Midwest Heavy Equipment Inc. is a major regional distributor of large construction equipment. The company was started by Oscar Thompson in the 1950s to service contractors who were building the interstate highway system. From the beginning, Midwest sold a wide range of equipment to contractors throughout the region. The company maintained a large stock of new and used equipment and established a reputation for reliably supplying contractors with their equipment needs. A large-scale service operation provided both on-site and shop-based repair services. In the beginning, many of the contractors were starting or expanding rapidly as they obtained a number of lucrative contracts.

Oscar Thompson established a reputation of support for new contractors, offering them quick access to equipment and services. As a result, these new companies did not need to buy various extra pieces of equipment that then sat idle while waiting for repair or for assignment to some incoming contract. Oscar also extended credit generously and did not pressure new firms for rapid payment. In the early days, he negotiated low-cost lines of credit with a number of large banks to cover the extended credit to his customers.

Because of its high service level to growing firms, Midwest did not need to compete on price. Midwest grew to be one of the largest suppliers of construction equipment in the region. The company employs over 3,000 people in its central facility in Minneapolis, and in fifty service and supply facilities from Ohio to Colorado and from Minnesota to Texas. Most contractors remember the high-quality service and attractive payment extensions provided by Oscar Thompson in the early years of their operation. Therefore, Midwest Equipment has had a large following of loyal customers. Midwest was a very profitable business through the 1960s and most of the 1970s. Oscar Thompson became a wealthy man and contributed generously to a number of colleges and other needs in the region.

In the mid-1970s, Oscar's daughter, Megan, completed her MBA at a top-ten school and began working with the company. Oscar had always planned that she would take over the company when he retired. During her college years she worked at a number of jobs, including mechanics helper, customer service representative, and financial operations associate. She often accompanied her father when he called on major customers, and she also drove a truck to deliver equipment. After obtaining her MBA, Megan worked in a number of marketing, financial, and management positions.

During the late 1970s and early 1980s, Midwest Heavy Equipment began to feel pressures similar to those experienced by a number of businesses. Foreign manufacturers of heavy equipment were setting up their own distribution facilities in direct competition with Midwest. These companies sold high-quality equipment at prices generally below those charged by Midwest. Contractors tended to be better established, and the competition between contractors was severe. Therefore, contractors sought to reduce costs any way they could. In addition, interest rates rose to high levels. This greatly increased the debt service cost of Midwest's line of credit, which it had traditionally used to cover inventory and accounts receivable.

12.2 TESTING FOR ERRORS IN REPORTED RECEIVABLES

Megan tells you she is concerned that, over the years, the number of errors in reporting receivables may have increased. Midwest has been using the same reporting procedure for the past twenty years, and the volume of receivables has grown by a factor of 5. Thus it may be necessary to design and install new reporting procedures for receivables. Several senior managers in the firm have advocated changing to a new computer information system that includes the reporting of receivables. As vice president of finance, Megan is also worried about both the dollar value of receivables and their length of time past due. She is considering various strategies for reducing the financial burden of receivables without damaging the company's relationship with loyal customers. Before developing detailed plans and taking action, she needs to understand the pattern of receivables.

After receiving your initial report on receivables, Megan asks you to audit the receivables for important characteristics. She has three important questions:

1. What percentage of receivables contain an error? This is the percentage of receivables that show a difference between the book value and the audited value.

2. What percentage of the receivables are more than 120 days past due?

3. What percentage of the receivables have a book value of $200,000 or more?

4. Are the errors in receivables related to the length of time for the receivables or to the dollar value of receivables?

You recommend that standard auditing attributes sampling procedures, which obtain percentage estimates based on a random sample of the receivables records, be used to answer these questions. To obtain the proper sample size, you must make preliminary estimates of expected population parameters and of the level of risk—probability of error—in the estimates.

A random sample of receivables will be used to describe the age and dollar value of receivables. The receivables records are contained in the data file named Midwest.[1]

12.3 RANDOM SAMPLES FOR AUDITING

After obtaining a description of accounts receivable, Megan decides to have you conduct a formal audit of receivables. This audit will examine the receivables and determine whether the level of reporting errors is excessive. The audit's results will be used to develop improved accounting procedures for receivables.

In this project you are to use standard auditing procedures to obtain and analyze a random sample of receivables. The procedures are included under the topic of attributes sampling, which is discussed in standard auditing textbooks. Attributes sampling uses a simple random sample for proportions with a normal approximation for the distribution of sample proportions. The analysis considers the "Deviation Rate," which is the percentage of individual receivables for which book value and audit value differ. Certain technical terms used by auditors need to be defined here and related to standard statistical procedures for simple random samples and one-tailed confidence intervals:

Expected Population Deviation Rate (EPDR): This is the expected rate (percentage) of deviations from the control in the population. For this project, the expected rate of deviations between book value and audit value in the population of all receivables is an example of the EPDR.

Tolerable Deviation Rate (TDR): This is the maximum rate (percentage) of deviation from the prescribed internal control procedure that can be tolerated in the population. In this project, it is the maximum percentage of deviations between book value and audit value that management is willing to tolerate before intervening. Thus it represents an upper control limit for the acceptance interval with the EPDR as the mean.

---

[1]The data for this study were generously supplied by Ernst & Young. Sincere thanks to Mark Hornung, senior manager of the Minneapolis office, and Mike Giese from the company's national office in Cleveland.

Confidence or Reliability Level: This is a judgmental decision relating to the percentage of time the sampling procedure yields an estimate that differs from the population by no more than the allowance for sampling error. Statistically, it is the probability α of exceeding the upper limit of the acceptance interval, with EPDR as the mean and TDR as the upper limit. For this project, the risk is set at $\alpha = 0.05$. This implies a standard normal Z_α value of 1.645.

Determination of Sample Size: The sample size is computed by using the following basic equation when sample proportions are estimated with a specified one-sided acceptance interval and a specified level of risk:

$$ n = \frac{Z_\alpha^2(P)(1-P)}{(\Delta P)^2} $$

where

P = Expected Population Deviation Rate (EPDR)

ΔP = Tolerable Deviation Rate (TDR − EPDR)

Z_α = 1.645

These definitions will be used in the questions at the end of this chapter. The variables to be used in this study are as follows:

Ident: Receivable record identification number.

Pstduday: Number of days the receivable is past its due date.

Bookval: Dollar value entered into the receivable record file.

Auditval: Actual dollar value as determined by the external auditors.

In accordance with Megan's wishes, you undertake the following steps.

12.1 Load the Midwest data file into your local computer system.

12.2 Obtain a random sample, and determine whether the percentage of errors exceeds the Tolerable Deviation Rate (TDR) of 4 percent. The Expected Population Deviation Rate (EPDR) is 2 percent. The EPDR is based on the historical error rate, which was actually last measured 10 years ago.

    a. Determine the sample size *n,* given a reliability level of 5 percent.

    b. Select a random sample of *n* receivables records from the Midwest data file. This can be accessed by using your local statistical system.

    c. Does the sample proportion exceed the TDR? Compute a two-sided confidence interval for the population deviation rate.

12.3 Obtain a random sample of records to determine the percentage of receivables that exceed $200,000, based on book value. The TDR for receivables exceeding $200,000 is 5 percent, and the EPDR is 3 percent. Megan believes that a high percentage of receivables with values above $200,000 would place a large interest cost on the company and would lower its most favorable credit rating.

   a. Determine the sample size n, given a reliability level of 5 percent.

   b. Select a random sample of n receivables records from the Midwest data file. This can be accessed by using your local statistical system.

   c. Does the sample proportion exceed the TDR? Compute a 95 percent two-sided confidence interval for the population deviation rate.

12.4 Obtain a random sample of records to determine the percentage of receivables that are more than 120 days past due. Megan has established a policy that no more than 20 percent of the receivables should be more than 120 days past due. This level is based in part on a strong comment by Midwest's principle bank. Based on historical operations, she estimates that the expected percentage of receivables more than 120 days past due (EPDR) is 16 percent.

   a. Determine the sample size n, given a reliability level of 5 percent.

   b. Select a random sample of n receivables records from the Midwest data file. This can be accessed by using your local statistical system.

   c. Does the sample proportion exceed the TDR? Compute a 95 percent two-sided confidence interval for the population deviation rate.

12.5 Prepare a one-page memo to the vice president of finance presenting your conclusions based on the sampling study. Include recommendations with regard to any problems you encountered with receivables, including the areas where action should be taken. Attach a one-page appendix showing how you computed your sample size for each of the three samples.

MIDWEST HEAVY EQUIPMENT INC. C

Differences Between Book and Audit Receivables

13.1 INTRODUCTION

Midwest Heavy Equipment Inc. is a major regional distributor of large construction equipment. The company was started by Oscar Thompson in the 1950s to service contractors who were building the interstate highway system. From the beginning, Midwest sold a wide range of equipment to contractors throughout the region. The company maintained a large stock of new and used equipment and established a reputation for reliably supplying contractors with their equipment needs. A large-scale service operation provided both on-site and shop-based repair services. In the beginning, many of the contractors were starting or expanding rapidly as they obtained a number of lucrative contracts.

Oscar Thompson established a reputation of support for new contractors, offering them quick access to equipment and services. As a result, these new companies did not need to buy various extra pieces of equipment that then sat idle while waiting for repair or for assignment to some incoming contract. Oscar also extended credit generously and did not pressure new firms for rapid payment. In the early days, he negotiated low-cost lines of credit with a number of large banks to cover the extended credit to his customers.

Because of its high service level to growing firms, Midwest did not need to compete on price. Midwest grew to be one of the largest suppliers of construction equipment in the region. The company employs over 3,000 people in its central facility in Minneapolis, and in fifty service and supply facilities from Ohio to Colorado and from Minnesota to Texas. Most contractors remember the high-quality service and attractive payment extensions provided by Oscar Thompson in the early years of their operation. Therefore, Midwest Equipment has had a large following of loyal customers. Midwest was a very profitable business through the 1960s and most of the 1970s. Oscar Thompson became a wealthy man and contributed generously to a number of colleges and other needs in the region.

In the mid-1970s, Oscar's daughter, Megan, completed her MBA at a top-ten school and began working with the company. Oscar had always planned that she would take over the company when he retired. During her college years she worked at a number of jobs, including mechanics helper, customer service representative, and financial operations associate. She often accompanied her father when he called on major customers, and she also drove a truck to deliver equipment. After obtaining her MBA, Megan worked in a number of marketing, financial, and management positions.

During the late 1970s and early 1980s, Midwest Heavy Equipment began to feel pressures similar to those experienced by a number of businesses. Foreign manufacturers of heavy equipment were setting up their own distribution facilities in direct competition with Midwest. These companies sold high-quality equipment at prices generally below those charged by Midwest. Contractors tended to be better established, and the competition between contractors was severe. Therefore, contractors sought to reduce costs any way they could. In addition, interest rates rose to high levels. This greatly increased the debt service cost of Midwest's line of credit, which it had traditionally used to cover inventory and accounts receivable.

13.2 DIFFERENCE BETWEEN BOOK AND AUDIT VALUE OF RECEIVABLES

In this case analysis of receivables, you are to determine whether a statistically significant difference exists between the book value and the audit value of receivables. Megan wishes to know whether the mean value of the audited receivables is less than the mean book value. If the audited mean values are consistently less, Midwest is overstating the value of its receivables and thus the net worth of the company. Overstatements can generate serious concerns among potential lenders and investors. In addition, Midwest has an ethical responsibility to its employees, its customers, and the community to report its true net worth. When he started the company, Oscar Thompson established the principle of complete honesty in both business and personal activities, and he encouraged this value in his daughter, too.

Based on your experience and the budget available for auditing receivables, you and Megan agree that a random sample of size $n = 200$ receivables will be used for the analysis. For your analysis you will obtain a random sample of size 200 from the Midwest data file, which you can access by using your statistical system. Comparisons will be made assuming both independent and dependent random samples of book value and audit value. You will be asked to recommend one of these two results and to indicate the rationale for your choice in the context of the problem.[1]

---

[1]The data for this study were generously supplied by Ernst & Young. Sincere thanks to Mark Hornung, senior manager of the Minneapolis office, and Mike Giese from the company's national office in Cleveland.

The variables to be used in this study are as follows

Ident: Receivable record identification number.

Pstduday: Number of days the receivable is past its due date.

Bookval: Dollar value entered into the receivable record file.

Auditval: Actual dollar value as determined by the external auditors.

In accordance with your discussions with Megan, you plan to take the following steps:

13.1 Load the Midwest data file on your local computer system.

13.2 Write the formal test that will provide strong evidence ($\alpha = 0.03$) that audit value is less than book value. Define the test statistic, its probability distribution, and the mean of the distribution.

13.3 Obtain a random sample of $n = 200$ receivables from the population in the Midwest file. This sample will be used for the remainder of the analysis.

13.4 Conduct the test using the assumption that the audit sample and the book value samples are independent. Assume that the two population variances are equal, and use the pooled estimator for the population standard deviations. State your conclusion in the language of statistical hypothesis testing.

13.5 Conduct the test using the assumption that the audit sample and the book value samples are dependent. Notice that there is an easy way and a hard way to include the covariance effect. State your conclusion in the language of statistical hypothesis testing.

13.6 Write a one- or-two-paragraph rationale indicating which of the preceding two testing procedures is appropriate. The rationale should be based on both the process by which the data were generated and the power of the resulting hypothesis test.

13.7 Write a one-page memo to the vice president of finance indicating the results of your analysis. Your report should be based on the preferred hypothesis testing procedure. Indicate the sample mean difference between book value and audit value, and state whether this difference is statistically significant or whether it could be the result of sampling error. Use your results to estimate the difference in total book value versus total audit value of receivables for the entire population of 1,661 receivables. Provide appropriate recommendations for improving the accounting system.

PRAIRIE FLOWER CEREAL INC. A

Cereal Packaging

14.1 INTRODUCTION AND BACKGROUND

Prairie Flower Cereal Inc. is a small but growing midwestern producer of hot and ready-to-eat breakfast cereals. The company was started in 1910 by Gordon Thorson, a successful grain farmer.[1] Originally it produced a cereal that, when cooked, was a basic breakfast food for many small-town midwestern families. In more recent years the company began to produce ready-to-eat cereals for the generic and special label market. The latter market has grown while the hot cereal market has declined. In 1990, Gordon's granddaughter Kristen took over as president replacing her father, Robert, who became chairman of the board.

Before becoming president, Kristen began her career with a large Fortune 500 producer of dry cereal and other food products. After eight years in various positions, including product manager, she moved to Prairie Flower Cereal and worked in a number of departments as part of her grooming to eventually take over the company. Kristen learned early that the consumer cereal business was a large and highly competitive business dominated by large companies that regularly introduced new products and implemented various promotions in attempts to capture market share. Before Kristen arrived, the company had introduced its own labels and sold a complete line of dry cereals at a lower price than the major brands. These occupied a small but stable niche, and the company was reasonably successful.

Prairie Flower Cereal has been growing rapidly in recent years. Its own brand label products have been very successful and the generic and private label business has shown steady growth. The corporate strategy has been to produce products of the same quality and taste as the major producers but at a selling price at least 25 percent lower. This strategy has been implemented by an efficient and modern production facility staffed with a dedicated work-

[1]The data for this case were obtained from the Malt-O-Meal Company, of Northfield, Minnesota. The assistance of John Stull, vice president of manufacturing, and of Cliff Okerlund, quality control, is gratefully acknowledged.

69

force. Advertising expenditures are small, which has provided an important cost advantage over the major producers. In addition the company packages its cereals in durable (but lower-cost) recyclable poly film bags instead of boxes. Prairie Flower's reputation has spread among a number of large supermarket chains. Consumers are initially attracted by the lower price but are then impressed by the quality and become regular purchasers. As a result, demand exceeds production capacity and a series of expansions has been planned. The company is reluctant to expand too rapidly, since it wishes to maintain the present close working relationship with all employees and effective management of all phases of the production process.

A high-quality dedicated workforce has been a key factor in the company's success. When Gordon Thorson began the firm, he developed a close "family" relationship with the employees in the original small factory. This close relationship has continued throughout the history of the firm. An employee profit-sharing program was developed a number of years ago. Production employees have received training from college accounting faculty so that they can understand the monthly accounting statements and their department's contribution to costs and profitability. Close personal relationships exist among workers at all levels from the executives to individual workers on the line.[2]

As a result of this long-term relationship with employees, the company has found it relatively easy to implement Demings 14 Points and other quality-control and production enhancement activities. An extensive education program has been implemented to increase the capabilities of all employees. Regular production meetings include production workers, quality-control staff, production management, and process engineering. These meetings have led to a number of improvements that have helped Prairie Flower maintain high quality and low unit cost. It is not unusual to hear production workers discussing the "sigma" of their process and identifying ways to reduce it.

14.2 CEREAL PACKING PROCESS IMPROVEMENT

You have been assigned to study the cereal-packing process and recommend ways to reduce costs. This process occurs at the end of the cereal production lines and fills the poly film bags with cereal to be sold in supermarkets. A schematic diagram of the production process is shown in Figure 14.1.

The process requires a balance between the requirement that bags be filled at or above the label weight and the requirement that they not be filled so

---

[2]Several years ago the company purchased a block of 20 tickets for each of the four home games the Minnesota Twins were to play in that year's World Series. These were distributed by a random lottery open to all employees. The manufacturing executive vice president thus reported to his friends at Rotary that he was not able to attend any of the World Series games because his name was not selected in the lottery.

| **FIGURE 14.1** | CEREAL PRODUCTION PROCESS |

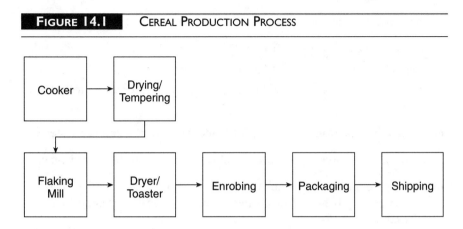

much above the label weight that the number of bags—production units—from each production run declines. Your study is to consider the performance of packing machines 1 and 2. A random sample of 100 observations was obtained from both machines while they were packing bags with a label weight of 1,134 grams or 40 ounces. The data are stored in a data file named PrairieA. Each observation is the gram weight of a randomly selected bag of cereal obtained during the production run. The weights from packing machine 1 are labeled Weight1 and are contained in the first column of the data file, and the weights from machine 2 are labeled Weight 2 and are contained in the second column of the data file.

Your study has a number of objectives. You are to determine the mean and the variance of the present machine settings. Historically, machine settings have been set so that no cereal bags weigh less than the label weight. Machine operators have operated with the belief that even the lowest weight package should be heavier than the label weight. Finally the production manager has asked you to consider the question of replacing either or both packing machines. The proposed new machine has a standard deviation of σ 2.0 grams when filling 40-ounce bags.

You begin your study by holding discussions with both the production and marketing managers, after which you all agree that the packing machine should be set so that no more than 1.5 bags per thousand fall below the label weight of 1,134 grams. You conclude that this level corresponds to the lower 3σ limit. Both machines underwent a complete regular maintenance and adjustment before the random sample of packages was obtained. Thus the machines were operating with their lowest possible standard deviation. Each 10 grams—approximately 1 percent—of extra cereal per package translates into an additional annual cost of $30,000 in lost revenue from the sale of cereal. A new packing machine would cost $150,000 including transportation, installation, and adjustment.

The variables contained in the PrairieA data file are as follows:

Weight1: Package weights from filling machine 1.
Weight2: Package weights from filling machine 2.

Consistent with your consultative discussions, you undertake the following steps.

14.1 Load the data from the PrairieA file into your local computer system.

14.2 Prepare an analysis to test the stability of package weights over time.

14.3 Compute the sample mean, standard deviation, range, quartiles, and other descriptive statistics for the sample from each package-filling machine.

14.4 Prepare a frequency distribution and histogram for the sample of package weights from each machine.

14.5 [Optional, if no further analysis is conducted.] Prepare a half-page report discussing the capability of each machine, based on the average and variability of package weights in the random sample.

14.6 For each machine, determine the minimum mean weight setting such that only 1.5 packages per thousand fall below the package label weight. Use the sample standard deviation as the estimate of the population standard deviation σ.

14.7 Compute the annual additional sales revenue for each machine that could be achieved by reducing the mean package weight from that observed in the sample to the minimum mean weight determined in the previous question.

14.8 Compute the annual additional sales revenue that would result if machine 1 were replaced by a new machine with a standard deviation of 2 grams, as compared to the standard deviation computed in this study.
    a. Compute the minimum mean setting for a lower 3σ at the label weight, using the sample standard deviation as an estimate of $\sigma >$
    b. Compute the minimum mean setting for a lower 3σ at the label weight, using the standard deviation of the proposed new machine.
    c. Compute the difference in mean weights or the grams of cereal that would be saved for each package.
    d. Using the annual savings per 10 grams of mean package weight reduction, compute the annual additional revenue that would be generated if machine 1 were replaced.

14.9   Compute the annual additional sales revenue that would result if machine 2 were replaced by a new machine with a standard deviation of 2 grams, as compared to the standard deviation computed in this study.

   a. Compute the minimum mean setting for a lower 3σ acceptance limit at the label weight, using the sample standard deviation as an estimate of σ.

   b. Compute the minimum mean setting for a lower 3σ at the label weight, using the proposed machine's standard deviation.

   c. Compute the difference in mean weights or the grams of cereal that would be saved for each package.

   d. Using the annual additional revenue per 10 grams of mean package weight reduction, compute the annual savings that would be generated if machine 2 were replaced.

14.10   Write a one-page report discussing the performance of the packaging machines and presenting your recommendations regarding the replacement of packing machines 1 and 2. Clearly indicate your rationale, including costs and benefits.

PRAIRIE FLOWER CEREAL INC. B

Analysis of Bulk Density

15.1 INTRODUCTION AND BACKGROUND

Prairie Flower Cereal Inc. is a small but growing midwestern producer of hot and ready-to-eat breakfast cereals. The company was started in 1910 by Gordon Thorson, a successful grain farmer.[1] Originally it produced a cereal that, when cooked, was a basic breakfast food for many small-town midwestern families. In more recent years the company began to produce ready-to-eat cereals for the generic and special label market. The latter market has grown while the hot cereal market has declined. In 1990, Gordon's granddaughter Kristen took over as president replacing her father, Robert, who became chairman of the board.

Before becoming president, Kristen began her career with a large Fortune 500 producer of dry cereal and other food products. After eight years in various positions, including product manager, she moved to Prairie Flower Cereal and worked in a number of departments as part of her grooming to eventually take over the company. Kristen learned early that the consumer cereal business was a large and highly competitive business dominated by large companies that regularly introduced new products and implemented various promotions in attempts to capture market share. Before Kristen arrived, the company had introduced its own labels and sold a complete line of dry cereals at a lower price than the major brands. These occupied a small but stable niche, and the company was reasonably successful.

Prairie Flower Cereal has been growing rapidly in recent years. Its own brand label products have been very successful and the generic and private label business has shown steady growth. The corporate strategy has been to produce products of the same quality and taste as the major producers but at a selling price at least 25 percent lower. This strategy has been implemented by an efficient and modern production facility staffed with a dedicated work-

[1]The data for this case were obtained from the Malt-O-Meal Company, of Northfield, Minnesota. The assistance of John Stull, vice president of manufacturing, and of Cliff Okerlund, quality control, is gratefully acknowledged.

force. Advertising expenditures are small, which has provided an important cost advantage over the major producers. In addition the company packages its cereals in durable (but lower-cost) recyclable poly film bags instead of boxes. Prairie Flower's reputation has spread among a number of large supermarket chains. Consumers are initially attracted by the lower price but are then impressed by the quality and become regular purchasers. As a result, demand exceeds production capacity and a series of expansions has been planned. The company is reluctant to expand too rapidly, since it wishes to maintain the present close working relationship with all employees and effective management of all phases of the production process.

A high-quality dedicated workforce has been a key factor in the company's success. When Gordon Thorson began the firm, he developed a close "family" relationship with the employees in the original small factory. This close relationship has continued throughout the history of the firm. An employee profit-sharing program was developed a number of years ago. Production employees have received training from college accounting faculty so that they can understand the monthly accounting statements and their department's contribution to costs and profitability. Close personal relationships exist among workers at all levels from the executives to individual workers on the line.[2]

As a result of this long-term relationship with employees, the company has found it relatively easy to implement Demings 14 Points and other quality-control and production enhancement activities. An extensive education program has been implemented to increase the capabilities of all employees. Regular production meetings include production workers, quality-control staff, production management, and process engineering. These meetings have led to a number of improvements that have helped Prairie Flower maintain high quality and low unit cost. It is not unusual to hear production workers discussing the "sigma" of their process and identifying ways to reduce it.

15.2 ANALYSIS OF CEREAL BULK DENSITY

The bulk density of cereal as it moves along the packing line is a critical measure of quality for the final product. A schematic diagram of the production process is shown in Figure 15.1.

During production, high density can occur either because the individual pieces of cereal are hard and brittle or because increased breakage has led to

---

[2]Several years ago the company purchased a block of 20 tickets for each of the four home games the Minnesota Twins were to play in that year's World Series. These were distributed by a random lottery open to all employees. The manufacturing executive vice president thus reported to his friends at Rotary that he was not able to attend any of the World Series games because his name was not selected in the lottery.

FIGURE 15.1   CEREAL PRODUCTION PROCESS

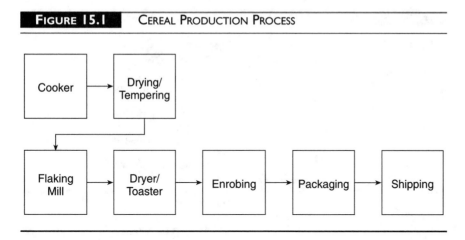

packing. Generally, a narrow range of bulk density provides a good taste for the consumer. Higher-density cereals are hard and crisp and hence crunchy when eaten. If they are too hard, chewing becomes excessively difficult, and this detracts from the consumer's enjoyment. In addition, high density may indicate breakage and powder, which are highly undesirable to consumers. Lower-density cereals tend to be puffy and munchy and easier to chew. However, if the density is too low the product becomes mushy and taste ratings drop sharply. Bulk density is also a concern for cereal packaging. Cereal is sold by weight; and if bulk density is too low, the required package volume may exceed the available package volume.

George Braken, the production manager, has asked you to study the bulk density for two cereal production lines. Line 1 is a higher-density flaked cereal, and line 2 is a lower-density "O" shaped cereal. The cereals are cooked, shaped, dried, toasted, sugar-coated, and then packed. George wants to know the central tendency and the variability of bulk density measures at three key points on the production line:

1. After drying and toasting
2. After sugar-coating (enrobing)
3. After being transported to the packaging machine.

After estimating present densities at those three locations on the production line, you are asked to estimate the change from one location to the next. George also wants to know whether the bulk density at intermediate points predicts bulk density at the packaging station.

The product currently receives strong positive evaluations from taste-test panels that evaluate randomly selected boxes of cereal. Thus George believes that the present range of bulk densities is consistent with a satisfactory

TABLE 15.1	VARIABLE NAMES IN THE PRAIRIEB DATA FILE	
VARIABLE NAME	**NUMBER OF OBSERVATIONS**	**VARIABLE DESCRIPTION**
Jetzone1	28	Bulk density (grams/liter), flaked cereal, after toaster
Sugarct1	28	Bulk density (grams/liter), flaked cereal, after suger coater
Packing1	28	Bulk density (grams/liter), flaked cereal, before packer
Jetzone2	28	Bulk density (grams/liter), "O" cereal, after toaster
Sugarct2	28	Bulk density (grams/liter), "O" cereal, after sugar coater
Packing2	28	Bulk density (grams/liter), "O" cereal, before packer

product taste image. He knows that competition is increasing, however, and anticipates that it will eventually become necessary to reduce the variance in bulk density. George wants to know the present distribution and potential strategies for future implementation to reduce variance.

Table 15.1 defines the variables that were measured for this study. A sample of twenty-eight observations was obtained from each production line at randomly selected time intervals and stored in a file named PrairieB. The bulk density measurements are recorded in grams per liter. Cereal 1 is a standard flaked cereal that is sugar-coated on this line. The three location samples were obtained at different times such that the same physical batch of cereal is being sampled for each observation. Cereal 2 is the company's popular "O's" cereal, which is coated with sugar and cinnamon flavor.

The planning group has asked you to analyze the bulk density at the various stations on the production line. After discussions with George and a number of Prairie Flower staffers, you conclude that the key tasks in this study include the following.

15.1 Load the data from the Prairie B file into your local computer system.

15.2 Prepare a time series plot of the bulk density measurements, and comment on their stability over time. [*Note:* You could use an *x*-bar process control chart, if that capability exists in your statistical package.]

15.3 Determine the distribution of bulk density at the measured points along the production line.

15.4 Calculate the increase in bulk density as the product moves between the various stations. Change in bulk density results from a combination of product change (such as drying or adding material) and product breakage.

15.5 Identify the predictability of final bulk density from previous measurements on the production line.

15.6 Establish acceptance intervals[3] for material in process, based on the following reference points.

    a. Symmetric intervals that exclude product having product density extremes. It is generally agreed that product flavor is equally degraded by either extremely low or extremely high bulk density.

    b. Minimum density for package filling. For example the 16-ounce or 454-gram plastic bag can contain a maximum of 0.60 liters. A finding that product density is too low implies that 16 ounces of the product will not fit into the bag.

15.7 Prepare a written report discussing the bulk density of the product at different points along the production line. Indicate how predictable bulk density at the packing machine is from measurements of bulk density at the other intermediate points.

---

[3]In this problem, acceptance intervals are established by using the 3σ confidence intervals from your analysis. Use the sample standard deviation as the estimate of σ. The process is assumed to operate "in control,' with a minimum variance. In addition, customer tests indicate that the product density is at the correct level.

PRAIRIE FLOWER CEREAL INC. C

Factors Influencing Bulk Density: Production Data

16.1   INTRODUCTION AND BACKGROUND

Prairie Flower Cereal Inc. is a small but growing midwestern producer of hot and ready-to-eat breakfast cereals. The company was started in 1910 by Gordon Thorson, a successful grain farmer.[1] Originally it produced a cereal that, when cooked, was a basic breakfast food for many small-town midwestern families. In more recent years the company began to produce ready-to-eat cereals for the generic and special label market. The latter market has grown while the hot cereal market has declined. In 1990, Gordon's granddaughter Kristen took over as president replacing her father, Robert, who became chairman of the board.

Before becoming president, Kristen began her career with a large Fortune 500 producer of dry cereal and other food products. After eight years in various positions, including product manager, she moved to Prairie Flower Cereal and worked in a number of departments as part of her grooming to eventually take over the company. Kristen learned early that the consumer cereal business was a large and highly competitive business dominated by large companies that regularly introduced new products and implemented various promotions in attempts to capture market share. Before Kristen arrived, the company had introduced its own labels and sold a complete line of dry cereals at a lower price than the major brands. These occupied a small but stable niche, and the company was reasonably successful.

Prairie Flower Cereal has been growing rapidly in recent years. Its own brand label products have been very successful and the generic and private label business has shown steady growth. The corporate strategy has been to produce products of the same quality and taste as the major producers but at a selling price at least 25 percent lower. This strategy has been implemented by an efficient and modern production facility staffed with a dedicated work-

---

[1]The data for this case were obtained from the Malt-O-Meal Company, of Northfield, Minnesota. The assistance of John Stull, vice president of manufacturing, and of Cliff Okerlund, quality control, is gratefully acknowledged.

force. Advertising expenditures are small, which has provided an important cost advantage over the major producers. In addition the company packages its cereals in durable (but lower-cost) recyclable poly film bags instead of boxes. Prairie Flower's reputation has spread among a number of large supermarket chains. Consumers are initially attracted by the lower price but are then impressed by the quality and become regular purchasers. As a result, demand exceeds production capacity and a series of expansions has been planned. The company is reluctant to expand too rapidly, since it wishes to maintain the present close working relationship with all employees and effective management of all phases of the production process.

A high-quality dedicated workforce has been a key factor in the company's success. When Gordon Thorson began the firm, he developed a close "family" relationship with the employees in the original small factory. This close relationship has continued throughout the history of the firm. An employee profit-sharing program was developed a number of years ago. Production employees have received training from college accounting faculty so that they can understand the monthly accounting statements and their department's contribution to costs and profitability. Close personal relationships exist among workers at all levels from the executives to individual workers on the line.[2]

As a result of this long-term relationship with employees, the company has found it relatively easy to implement Demings 14 Points and other quality-control and production enhancement activities. An extensive education program has been implemented to increase the capabilities of all employees. Regular production meetings include production workers, quality-control staff, production management, and process engineering. These meetings have led to a number of improvements that have helped Prairie Flower maintain high quality and low unit cost. It is not unusual to hear production workers discussing the "sigma" of their process and identifying ways to reduce it.

16.2 ANALYSIS OF CEREAL BULK DENSITY

The bulk density of cereal as it moves along the packing line is a critical measure of quality for the final product. A schematic diagram of the production process is shown in Figure 16.1.

During production, high density can occur either because the individual pieces of cereal are hard and brittle or because an unusually high level of

[2]Several years ago the company purchased a block of 20 tickets for each of the four home games the Minnesota Twins were to play in that year's World Series. These were distributed by a random lottery open to all employees. The manufacturing executive vice president thus reported to his friends at Rotary that he was not able to attend any of the World Series games because his name was not selected in the lottery.

FIGURE 16.1   CEREAL PRODUCTION PROCESS

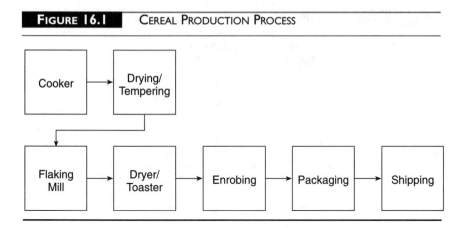

breakage and consequent packing has occurred. Generally only a narrow range of bulk density provides a good taste for the consumer. Higher-density cereals are hard and crisp and hence crunchy to the taste. If they are too hard, chewing becomes more difficult and this detracts from taste. In addition, high density may also indicate breakage and powder, which are very undesirable for the consumer. Lower-density cereals tend to be puffy and munchy and easier to chew. If the density is too low, however, the product is mushy and taste is poor. Bulk density is also an important concern for cereal packaging. Cereal is sold by weight; and if bulk density is too low, the required package volume may exceed the available package volume.

George Braken, production manager, has asked you to prepare a study of the dryer/toaster portion of the production line. He wants to know how the operating variables of this process influence the bulk density of the resulting cereal. Percent moisture and bulk density are thought to be strongly related because drying generally reduces density. Therefore, at the same time, he wants you to determine the factors that influence percent moisture.

The dryer/toaster is essentially a long oven that processes cereal passing through it on a slow-moving conveyer belt. This process follows the forming process, during which the cereal is formed into its desired shape by the rollers or extruders. Drying and toasting result in strong stable flakes or "O's," which are then packed and delivered to retail stores. The cereal passes through two zones of the dryer/toaster on its conveyor belt. The first zone has a higher temperature, which produces a hardened surface without excessive drying. The second zone has a lower temperature. In this stage, the cooking and drying continue, contributing to a high-quality product. The quality of the cereal produced is controlled by the temperature inside the toaster and by the fluidity of the cereal drying bed. Cereal bed fluidity is controlled by air flow and other conditions inside the toaster. The combination of these conditions is expressed as a variable named Mag1 or Mag2, depending on the

TABLE 16.1	VARIABLE NAMES IN THE PRAIRIEC DATA FILE	

VARIABLE NAME	NUMBER OF OBSERVATIONS	VARIABLE DESCRIPTION
Row	44	Row or observation number
Moisture	44	Percent moisture
Bulkdens	44	Density of cereal mix (grams/liter)
Zone1	44	Temperature in toaster zone 1 (degrees F)
Zone2	44	Temperature in toaster zone 2 (degrees F)
Rate	44	Speed of conveyor (feet per minute)
Conveyor	44	Identification number for conveyor
Mag1	44	Fluidization measure in toaster zone 1
Mag2	44	Fluidization measure in toaster zone 2

zone. Density and its related measures must be carefully controlled to ensure a high-quality cereal that is pleasing to customers.

The quality of the resulting product is measured primarily by the bulk density and moisture content of the cereal after it leaves the dryer/toaster. In addition, experienced process analysts judge the product by its color on an ordinal scale: higher numbers indicate darker colors. By monitoring these measures, the company can maintain the quality of cereal at a high level to ensure customer satisfaction.

George has provided you with a set of data that was collected at various times from one of the production lines. These data are contained in a file named PrairieC and are described in Table 16.1. You tell George that the results of your statistical study would be much better if you could develop an experimental design and collect data at carefully defined operating levels. However, you agree to conduct a preliminary analysis using the data on hand.

After discussions with George and other members of the production staff, you identify the likely predictor variables. Most of the operators believe that the temperatures and the fluidity of the cereal bed in each of the zones greatly influence bulk density and percent moisture. In addition, the line speed—which determines the length of time a batch of cereal spends in the dryer/toaster—is also considered important. You decide to use multiple-regression analysis to select the important predictor variables and to estimate two linear prediction models. One model predicts bulk density and the second predicts percent moisture.

You are to prepare a short report that documents the conclusions of your analysis. This report should present your final models and indicate the variables that predict bulk density and moisture; it should also provide a description of the data. Include a discussion of any problems that result from using these data for your analysis. Indicate why data obtained by using an experimental design model are likely to provide better results.

16.1 Load the data from the PrairieC file into your local computer system.

16.2 Prepare *x*-bar and pi charts (sample size = 3) for percent moisture and product density. Discuss the stability of these measures over time.

16.3 Use your statistical package to prepare descriptive statistics and graphical descriptions for each of the variables in the data set. In addition, compute the matrix of simple correlations between the various variables.

16.4 Select potential predictor variables for the linear regression models that will predict bulk density and percent moisture.

16.5 Use multiple regression to determine which of the potential predictor variables are actually related to bulk density and to percent moisture. Based on this analysis, indicate the best variables to use to adjust percent moisture and bulk density.

16.6 Discuss the problems with the data used in this project. Indicate how these problems could lead to misleading results. Your answer should be specific.

16.7 Prepare a written report indicating the influence of dryer/toaster variables on product density and percent moisture. Discuss the analysis, and present key conclusions.

PRAIRIE FLOWER CEREAL INC. D

Factors Influencing Bulk Density: Experimental Design Data

17.1 INTRODUCTION AND BACKGROUND

Prairie Flower Cereal Inc. is a small but growing midwestern producer of hot and ready-to-eat breakfast cereals. The company was started in 1910 by Gordon Thorson, a successful grain farmer.[1] Originally it produced a cereal that, when cooked, was a basic breakfast food for many small-town midwestern families. In more recent years the company began to produce ready-to-eat cereals for the generic and special label market. The latter market has grown while the hot cereal market has declined. In 1990, Gordon's granddaughter Kristen took over as president replacing her father, Robert, who became chairman of the board.

Before becoming president, Kristen began her career with a large Fortune 500 producer of dry cereal and other food products. After eight years in various positions, including product manager, she moved to Prairie Flower Cereal and worked in a number of departments as part of her grooming to eventually take over the company. Kristen learned early that the consumer cereal business was a large and highly competitive business dominated by large companies that regularly introduced new products and implemented various promotions in attempts to capture market share. Before Kristen arrived, the company had introduced its own labels and sold a complete line of dry cereals at a lower price than the major brands. These occupied a small but stable niche, and the company was reasonably successful.

Prairie Flower Cereal has been growing rapidly in recent years. Its own brand label products have been very successful and the generic and private label business has shown steady growth. The corporate strategy has been to produce products of the same quality and taste as the major producers but at a selling price at least 25 percent lower. This strategy has been implemented by an efficient and modern production facility staffed with a dedicated work-

---

[1]The data for this case were obtained from the Malt-O-Meal Company, of Northfield, Minnesota. The assistance of John Stull, vice president of manufacturing, and of Cliff Okerlund, quality control, is gratefully acknowledged.

force. Advertising expenditures are small, which has provided an important cost advantage over the major producers. In addition the company packages its cereals in durable (but lower-cost) recyclable poly film bags instead of boxes. Prairie Flower's reputation has spread among a number of large supermarket chains. Consumers are initially attracted by the lower price but are then impressed by the quality and become regular purchasers. As a result, demand exceeds production capacity and a series of expansions has been planned. The company is reluctant to expand too rapidly, since it wishes to maintain the present close working relationship with all employees and effective management of all phases of the production process.

A high-quality dedicated workforce has been a key factor in the company's success. When Gordon Thorson began the firm, he developed a close "family" relationship with the employees in the original small factory. This close relationship has continued throughout the history of the firm. An employee profit-sharing program was developed a number of years ago. Production employees have received training from college accounting faculty so that they can understand the monthly accounting statements and their department's contribution to costs and profitability. Close personal relationships exist among workers at all levels from the executives to individual workers on the line.[2]

As a result of this long-term relationship with employees, the company has found it relatively easy to implement Demings 14 Points and other quality-control and production enhancement activities. An extensive education program has been implemented to increase the capabilities of all employees. Regular production meetings include production workers, quality-control staff, production management, and process engineering. These meetings have led to a number of improvements that have helped Prairie Flower maintain high quality and low unit cost. It is not unusual to hear production workers discussing the "sigma" of their process and identifying ways to reduce it.

17.2 ANALYSIS OF CEREAL BULK DENSITY

The bulk density of cereal as it moves along the packing line is a critical measure of quality for the final product. A schematic diagram of the production process is shown in Figure 17.1.

During production, high density can occur either because the individual pieces of cereal are hard and brittle or because an unusually high level of breakage and consequent packing has occurred. Generally only a narrow

---

[2]Several years ago the company purchased a block of 20 tickets for each of the four home games the Minnesota Twins were to play in that year's World Series. These were distributed by a random lottery open to all employees. The manufacturing executive vice president thus reported to his friends at Rotary that he was not able to attend any of the World Series games because his name was not selected in the lottery.

FIGURE 17.1   CEREAL PRODUCTION PROCESS

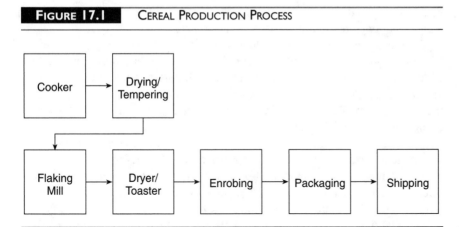

range of bulk density provides a good taste for the consumer. Higher-density cereals are hard and crisp and hence crunchy to the taste. If they are too hard, chewing becomes more difficult and this detracts from taste. In addition, high density may also indicate breakage and powder, which are very undesirable for the consumer. Lower-density cereals tend to be puffy and munchy and easier to chew. If the density is too low, however, the product is mushy and taste is poor. Bulk density is also an important concern for cereal packaging. Cereal is sold by weight; and if bulk density is too low, the required package volume may exceed the available package volume.

George Braken, production manager, has asked you to prepare a study of the dryer/toaster portion of the production line. He wants to know how the operating variables of this process influence the bulk density of the resulting cereal. Percent moisture and bulk density are thought to be strongly related because drying generally reduces density. Therefore, at the same time, he wants you to determine the factors that influence percent moisture.

The dryer/toaster is essentially a long oven that processes cereal passing through it on a slow-moving conveyer belt. This process follows the forming process, during which the cereal is formed into its desired shape by the rollers or extruders. Drying and toasting result in strong stable flakes or "O's," which are then packed and delivered to retail stores. The cereal passes through two zones of the dryer/toaster on its conveyor belt. The first zone has a higher temperature, which produces a hardened surface without excessive drying. The second zone has a lower temperature. In this stage, the cooking and drying continue, contributing to a high-quality product. The quality of the cereal produced is controlled by the temperature inside the toaster and by the fluidity of the cereal drying bed. Cereal bed fluidity is controlled by air flow and other conditions inside the toaster. The combination of these conditions is expressed as a variable named Mag1 or Mag2, depending on the zone. Density and its related measures must be carefully controlled to ensure a high-quality cereal that is pleasing to customers.

The quality of the resulting product is measured primarily by the bulk density and moisture content of the cereal after it leaves the dryer/toaster. In addition, experienced process analysts judge the product by its color on an ordinal scale: higher numbers indicate darker colors. By monitoring, these measures the company can maintain the quality of cereal at a high level to ensure customer satisfaction.

After your initial introduction to the problem, you recommend that a series of production experiments—part of an experimental design model—be conducted. You point out that a well-designed experiment will provide more information for each observation. This is important because Prairie Flower wishes to use the results to help control the dryer/toaster and thus produce a higher-quality product. You also note, however, that the dryer/toaster process must be run at non-standard and extreme levels in order to obtain good results from the experimental design. As a result, production needs may sometimes have to be compromised briefly so that the proper data points can be obtained. After some discussion, George agrees that a series of production experiments should be conducted, and he requests that you prepare the design.

The experimental design will be used to identify the effect on bulk density and moisture of the five major variables in the dryer/toaster, each of which will be set at two different levels:

Zone 1 temperature (400°F and 490°F)

Zone 2 temperature (290°F and 380°F)

Conveyor (no. 55 and no. 75)

Mag1, bed fluidity in zone 1 (2.0 and 3.2)

Mag2, bed fluidity in zone 2 (1.0 and 2.6)

To reduce the number of experimental observations, you decide to design a one-half factorial experiment, with only sixteen observations. If the full factorial experiment—which would include all combinations of the five variables—had been conducted a total of thirty-two experimental observations would have been required. The data from the experiment are contained in the PrairieD file and the variables are defined in Table 17.1. This design assumes that all relationships between the independent or design variables and the outcome or dependent variable are linear

You are to analyze the data from the experimental design and prepare a report that indicates how the variables influence bulk density and percent moisture. Your analysis can be carried either by using the original data or by converting the data to a series of 0 and 1 dummy variables. In both cases, multiple regression can be used to obtain the coefficients for the analysis

TABLE 17.1	VARIABLE NAMES IN THE PRAIRIED DATA FILE

VARIABLE NAME	NUMBER OF OBSERVATIONS	VARIABLE DESCRIPTION
Row	16	Row or observation number
Zone1	16	Temperature in toaster zone 1 (degrees F)
Zone2	16	Temperature in toaster zone 2 (degrees F)
Converyor	16	Converyor identification number
Mag1	16	Zone 1 fluidity of cereal bed on conveyor
Mag2	16	Zone 2 fluidity of cereal bed on conveyor
Moisture	16	Percent moisture after dryer/toaster
Bulkdens	16	Bulk density (grams/liter) after dryer/toaster
Color	16	Color rating after dryer/toaster

model. If the original data are used, the coefficients indicate the change in the dependent variable—bulk density or percent moisture—for each unit change of the independent variable. Alternatively, a dummy variable model provides estimates of the total change from the lower to the upper level of the independent variable—for instance, what the change in bulk density is when zone 1 temperature is increased from 400 to 490 degrees Fahrenheit. This later approach yields the same results as the classical experimental design applying a version of analysis of variance.

17.1 Prepare descriptive statistics for each of the variables in the data file, including means, variances, and correlations. Prepare a short written description of the variables, with special emphasis on the results from the correlation analysis.

17.2 Using multiple regression, fit a model that predicts bulk density as a function of the five design variables.

17.3 Fit a model that predicts percent moisture as a function of the design variables.

17.4 Convert the design variables to dummy variables, and use multiple regression to estimate the effects of the changes in the design variables on bulk density and percent moisture.

17.5 Prepare a report discussing the results of your analysis. Indicate which design variables have a significant effect on bulk density and percent moisture. Provide an estimate of the effect of those design variables.

RIVER VALLEY
HEALTH INSURANCE COMPANY A

Maternity Claims Analysis

18.1 INTRODUCTION

River Valley Health Insurance Company is a major supplier of health insurance in the upper Midwest, with over 400,000 policy holders. It is the oldest of the major insurers, having celebrated its sixtieth anniversary in 1992. Over the years its business has changed from emphasizing individual family policies to predominantly serving large corporate and government customers who purchase health insurance as a fringe benefit for their employees. River Valley also processes claims for government Medicaid and Medicare programs. During the past fifteen years, health-care costs have risen dramatically, and River Valley has been under significant pressure to reduce its operating costs and to negotiate lower payments for health-care providers (including medical doctors, hospitals, and drug suppliers).

River Valley Health Insurance has been a long-time advocate of high-quality health care, and it does not want lower provider payments to reduce the quality of health care its policy-holders receive. Instead the company wants to encourage selection of appropriate minimum-cost procedures, discourage excessive resources, and encourage efficient use of available resources.

To help manage the dual goals of low cost and high quality, River Valley has developed a sophisticated insurance claims processing system that runs on a large mainframe computer. That system checks to confirm that appropriate medical procedures were used to handle a given diagnosis. In addition, charges are compared with "usual and customary charges" to ensure that they do not exceed standard boundaries. This processing system also provides a large database of information that can be used to analyze costs for various medical procedures and patient groups. It has also been used to identify specific hospitals and geographic regions that experience unusually frequent use of a specific procedure or a very high occurrence of a particular disease. Several years ago analysis of these data revealed that an excessive number of people had developed heart disease in a small isolated town. As a result a major program was developed to reduce heart disease in that town. The program included education, improved diets, exercise programs, and routine examinations for all citizens over age 40.

Insurance companies such as River Valley have become a major force in controlling the rate of increase of health-care claims costs. By reducing claims costs, River Valley can charge lower premiums to policy holders. Increasingly, health insurance is provided as a standard fringe benefit by most large employers. However, as health-care costs have risen, the cost of providing insurance for employees has shot up dramatically. Several years ago General Motors reported that it spent more money on health-care insurance per car produced than it spent on the steel contained in the cars. As a result, large firms have increasingly tried to reduce health insurance costs. They have required competitive bids from insurance companies, and in some cases they have become partially self-insured and/or required employees to pay a larger share of the health insurance premium.

Health insurance companies have been forced to do all they can to reduce the amount they pay to health-care providers. For example, many insurers use their large buying power to negotiate lower payments to doctors and hospitals, by establishing groups of preferred providers who are eligible to provide services for the large number of health insurance policy-holders controlled by large insurance companies. Preferred providers are selected on the basis of their track record for providing good-quality health care at relatively low cost. Lower payment rates are negotiated with the selected providers, who are then monitored to ensure that they do not sacrifice high-quality patient care to achieve lower costs. Providers who do not agree to the reduced fees become ineligible to provide care for patients from large corporations and other groups established by the insurance companies.

18.2 ANALYSIS OF MATERNITY CLAIMS

River Valley Health Insurance's actuarial research division, directed by Charlotte Baldwin, conducts a number of analyses of the insurance claims data collected by the system. Analyses ultimately have a single objective: reducing health-care claims costs while maintaining high-quality medical care. The large database of insurance claims provides a rich resource for statistical analyses. Analysis results are used to identify low- and high-cost providers. In addition, the frequency of various medical procedures is tracked by geographic and demographic group, and overutilization of certain procedures by specific providers is identified. Typical costs are established from the large base of experience. All of these results are used in negotiations to establish the fees that will actually be paid to providers. In addition, providers with lower-cost experience per patient in various categories are most likely to be included in various pools.

TABLE 18.1	VARIABLES CONTAINED IN THE HEALTH DATA FILE
VARIABLE NAME	**VARIABLE DESCRIPTION**
Area	Geographic region (Rural area = 0; Urban area = 1)
Deliv	Type of delivery (Normal = 0; Caesarean = 1)
Mbrno	Sequential number identifying each woman patient
Age	Age of the patient at the time of delivery
Prodtype	Product (Primary care = 0; Fee for service = 1)
Matprfcg	Total maternity charges for professional services
Matfaccg	Total maternity charges for facility usage
Nomatdcg	Total charges for drugs during pregnancy period
Nomatocg	Total charges for nonmaternity procedures during pregnancy period
Submatp	Total subscriber-eligible charges for maternity professional services
Submatf	Total subscriber-eligible charges for maternity facilities services
Subdrug	Total subscriber-eligible charges for drugs during pregnancy period
Subother	Total subscriber-eligible charges for nonmaternity procedures during pregnancy period

As a member of the actuarial research group, you have been assigned to determine the factors that affect maternity claims costs. The analysis team includes Alice Miller, a systems analyst, who has primary responsibility for preparing a random sample of patient claims that will be used for the analysis.[1] The patient claims cost for the mother covers the period beginning nine months before the birth and extending until the mother is discharged from the hospital. After considerable discussion with Alice and the executive vice-president, Richard Lemers, you develop a list of appropriate variables and a selection strategy. You decide to obtain approximately the same number of claims from rural and from urban areas. Insurance claims are separated by professional and hospital facility categories, by the amount charges, and by the amount actually paid by River Valley Insurance. Separate categories of claims for drugs and for nonmaternity claims during the time period are collected. Table 18.1 contains a list of the variables in the data file named Health.

In developing the study, you are asked to identify the effect of certain factors. In particular, Richard wants to know how much additional cost is associated with a caesarean section delivery compared to a normal delivery, and he wants to know whether larger charges were submitted by urban hospitals and physicians than by rural ones. The company has two groups of policy

[1]The data for this case were generously supplied by Blue Cross/Blue Shield of Minnesota. The cooperation and assistance of Richard Niemic, senior vice president for corporate affairs, and Alice Miller, senior consultant, are gratefully acknowledged.

holders: standard "fee for service" insurance, and "primary care clinics" (such as health maintenance organizations). The latter are given a standard allowance for each client and are asked to control costs through careful selection of procedures and of specialist providers. The argument is that efficient medical decisions will be made because physicians who know the patient are making those decisions. Finally, Richard asked the team to determine whether the average charge differed for mothers of different ages.

After further discussion with the study team, you agree to prepare a statistical description of each of the four cost categories. You also undertake to determine whether the rate of occurrence of caesarean section procedures differs between rural and urban areas, whether it differs between primary-care and fee-for-service clinics, and whether it is significantly influenced by the mother's age. Finally you will determine whether differences in charges exist for urban versus rural, for fee-for-service versus primary-care clinics, for mothers of different ages, and for caesarean section versus normal delivery. The team also concurs that it is important to identify excessively large and small charges so that the claim records can be checked—either to identify procedural errors in the claims processing or to provide an understanding of unusual cases whose charges deviate greatly from typical cases.

You will prepare statistical analyses for the team and present the results at a future team meeting, and you will prepare a written report presenting and discussing your results. Your analysis should include the following steps.

18.1 Load the data from the Health file into your local computer system. (This is a large data file compared to those used in the other cases.)

18.2 Prepare a descriptive analysis of the variables in the data file.

18.3 Identify difference in the rate of occurrence of caesarean section procedures in rural versus urban, primary care versus fee for service, and by mother's age.

18.4 Identify differences in billed charges by caesarean section versus normal delivery, by rural versus urban, and by primary care versus fee for service.

18.5 Prepare a written report describing your analysis and presenting the results. Indicate recommendations for cost reductions.

RIVER VALLEY
HEALTH INSURANCE COMPANY B

Cost Reduction

19.1 INTRODUCTION

River Valley Health Insurance Company is a major supplier of health insurance in the upper Midwest, with over 400,000 policy holders. It is the oldest of the major insurers, having celebrated its sixtieth anniversary in 1992. Over the years its business has changed from emphasizing individual family policies to predominantly serving large corporate and government customers who purchase health insurance as a fringe benefit for their employees. River Valley also processes claims for government Medicaid and Medicare programs. During the past fifteen years, health-care costs have risen dramatically, and River Valley has been under significant pressure to reduce its operating costs and to negotiate lower payments for health-care providers (including medical doctors, hospitals, and drug suppliers).

River Valley Health Insurance has been a long-time advocate of high-quality health care, and it does not want lower provider payments to reduce the quality of health care its policy-holders receive. Instead the company wants to encourage selection of appropriate minimum-cost procedures, discourage excessive resources, and encourage efficient use of available resources.

To help manage the dual goals of low cost and high quality, River Valley has developed a sophisticated insurance claims processing system that runs on a large mainframe computer. That system checks to confirm that appropriate medical procedures were used to handle a given diagnosis. In addition, charges are compared with "usual and customary charges" to ensure that they do no exceed standard boundaries. This processing system also provides a large database of information that can be used to analyze costs for various medical procedures and patient groups. It has also been used to identify specific hospitals and geographic regions that experience unusually frequent use of a specific procedure or a very high occurrence of a particular disease. Several years ago analysis of these data revealed that an excessive number of people had developed heart disease in a small isolated town. As a result a major program was developed to reduce heart disease in that town. The program included education, improved diets, exercise programs, and routine examinations for all citizens over age 40.

Insurance companies such as River Valley have become a major force in controlling the rate of increase of health-care claims costs. By reducing claims costs, River Valley can charge lower premiums to policy holders. Increasingly, health insurance is provided as a standard fringe benefit by most large employers. However, as health-care costs have risen, the cost of providing insurance for employees has shot up dramatically. Several years ago General Motors reported that it spent more money on health-care insurance per car produced than it spent on the steel contained in the cars. As a result, large firms have increasingly tried to reduce health insurance costs. They have required competitive bids from insurance companies, and in some cases they have become partially self-insured and/or required employees to pay a larger share of the health insurance premium.

Health insurance companies have been forced to do all they can to reduce the amount they pay to health-care providers. For example, many insurers use their large buying power to negotiate lower payments to doctors and hospitals, by establishing groups of preferred providers who are eligible to provide services for the large number of health insurance policy-holders controlled by large insurance companies. Preferred providers are selected on the basis of their track record for providing good-quality health care at relatively low cost. Lower payment rates are negotiated with the selected providers, who are then monitored to ensure that they do not sacrifice high-quality patient care to achieve lower costs. Providers who do not agree to the reduced fees become ineligible to provide care for patients from large corporations and other groups established by the insurance companies.

19.2 COST REDUCTION ANALYSIS

River Valley Health Insurance's actuarial research division, directed by Charlotte Baldwin, conducts a number of analyses of the insurance claims data collected by the system. Analyses ultimately have a single objective: reducing health-care claims costs while maintaining high-quality medical care. The large database of insurance claims provides a rich resource for statistical analyses. Analysis results are used to identify low- and high-cost providers. In addition, the frequency of various medical procedures is tracked by geographic and demographic group, and overutilization of certain procedures by specific providers is identified. Typical costs are established from the large base of experience. All of these results are used in negotiations to establish the fees that will actually be paid to providers. In addition, providers with lower-cost experience per patient in various categories are most likely to be included in various pools.

TABLE 19.1	VARIABLES CONTAINED IN THE HEALTH DATA FILE

VARIABLE NAME	**VARIABLE DESCRIPTION**
Area	Geographic region (Rural area = 0; Urban area = 1)
Deliv	Type of delivery (Normal = 0; Caesarean = 1)
Mbrno	Sequential number identifying each woman patient
Age	Age of the patient at the time of delivery
Prodtype	Product (Primary care = 0; Fee for service = 1)
Matprfcg	Total maternity charges for professional services
Matfaccg	Total maternity charges for facility usage
Nomatdcg	Total charges for drugs during pregnancy period
Nomatocg	Total charges for nonmaternity procedures during pregnancy period
Submatp	Total subscriber-eligible charges for maternity professional services
Submatf	Total subscriber-eligible charges for maternity facilities services
Subdrug	Total subscriber-eligible charges for drugs during pregnancy period
Subother	Total subscriber-eligible charges for nonmaternity procedures during pregnancy period

River Valley Health Insurance's president, Pamela Smith, has asked for an evaluation of the program to establish preferred providers. She is particularly interested in the cost reductions that have been achieved through negotiations with provider groups.

You have been assigned to the team responsible for determining cost reductions for a limited set of medical procedure subgroups. The team consists of Alice Miller, systems analyst; George Barker, marketing analyst; Peggy Johnson, provider negotiations; and your team of statistical analysts. Your first project is to examine the effectiveness of cost containment for maternity-related claims.[1]

After some discussion, the team decides that the evaluation should be based on a stratified random sample of maternity cases that represent urban and rural providers equally. Peggy Johnson emphasizes that negotiations have tended to focus on urban clinics and hospitals because their charges were typically higher than those of their rural counterparts. Alice Miller agrees to obtain a random sample of approximately 2,500 maternity-related claims from the past year. The team agrees that, in addition to determining the overall effectiveness of claims cost reduction, the analysis should determine whether the reduction is different for urban versus rural providers and for fee-for-service versus primary-care clinic providers. Because of the large

[1]The data for this case were generously supplied by Blue Cross/Blue Shield of Minnesota. The cooperation and asistance of Richard Niemic, senior vice president for corporate affairs, and Alice Miller, senior consultant, are gratefully acknowledged.

cost difference involved, caesarean section claims costs will be separated from normal claims. Several team members also wonder whether patient age has any relationship to the cost reduction.

The variables in the data file obtained by Alice are described in Table 19.1. These data are in a file named Health. The charges billed and the subscriber-eligible charges—the amount actually paid to the provider—are divided into four subsets:

Maternity charges for professional services

Maternity charges for hospital facilities

Charges for drugs during the nine-month pregnancy period

Nonmaternity medical charges during the pregnancy period

The first two charges are usually the largest components of the bill. However, the group does wonder whether nonmaternity charges are related to complications that in turn result in large maternity-related charges. The team also agrees that claims with excessively high charges and payments should be identified for further review. Such review might indicate additional issues that should be included in provider negotiations. In addition, potential errors in claims processing might be identified.

After the meeting you agree that your team of statistical analysts will perform the analyses defined by the study team. You will prepare a written report discussing your results, and you will make a presentation at a future meeting. The analysis should include the following steps.

19.1 Load the data from the Health file into your local computer system.

19.2 Prepare a description of the variables in the data file.

19.3 Compute differences between the billed charges and the eligible charges, and prepare an analysis of the characteristics of these differences.

19.4 Develop a model to predict the reduction in charges as a function of the variables in the data file.

19.5 Prepare a written report describing your study and detailing the results regarding the reduction in claims. Provide recommendations about possible opportunities for future cost reductions based on your analysis.

STATE LOTTERY A

Description of Lottery Sales Performance

20.1 BACKGROUND

In recent years, voters have expressed a strong desire for lower taxes—without, however, expressing a commensurate "demand" for reduced or eliminated services. In partial response to this problem, a number of state legislatures have passed laws to establish a state lottery. The rationale is simple: people like to gamble; legal state-run gambling can be controlled to avoid negative side effects; and the state should take advantage of this situation to raise revenue without having to raise taxes.[1]

The State of Loonland recently established a state lottery and has hired an experienced lottery executive, Gerald O'Leary, to install and operate it. As part of the development plan, Gerry set up a data collection system to record sales by county. This system permits him to monitor the lottery operation from start-up onward. The lottery operation consists of selling, at various retail outlets, game cards that offer the possibility of winning an instant prize, with a specified probability. In addition, card numbers are used for a random selection of winners of much larger daily and weekly prizes. If no winner is identified for a daily or weekly prize, the prize pool grows, providing an opportunity for publicity to increase excitement and sales among the pool of prospective players. The probabilities associated with winning various prizes are established so that the expected return to players is 50 percent of total sales revenue. Administration of the state lottery apparatus consumes 10 percent of revenue, and the retail outlets receive 5 percent. Thus, 35 percent of total sales revenue is delivered to the treasury of Loonland. In the original bill establishing the lottery, this money was earmarked for conservation, pollution abatement, and education. In addition, a small amount—approximately

---

[1]Many states have developed state-run lotteries, so the legitimacy of this activity is not at issue here. Our concern is with the use of statistical analysis to assist in the managing of a state-run business. This case is based on real data, but all proper names are fictitious.

0.5 percent of revenue—has been designated for treatment of compulsive gamblers.

Gerald O'Leary has established a staff of strong professionals to develop and administer the new lottery. As time goes on, the lottery staff will continually devise new lottery game formats and will prepare advertising and promotional material to maintain a high level of interest among the pool of regular players. The actual game cards and the procedures for assigning the probability of winning are purchased from a supplier who guarantees the appropriate probabilities for the various prizes. Staff members also negotiate with retailers who apply to be sales outlets. The lottery staff trains store personnel to use equipment that issues the cards in each store. After a retailer has signed a contract as an authorized lottery sales outlet, the lottery staff regularly monitors each outlet to ensure good relations with players and correct reporting of revenue. Gerry has made his staff aware of the importance of good relationships between players and retailers. In addition, lottery administrators have developed procedures for monitoring sales to ensure proper reporting of sales revenue from each retail outlet.

20.2 MANAGEMENT ISSUES AND OPERATING DATA

Gerry knows the importance of good statistical analysis for monitoring the lottery start-up and its continued operation. He has asked you to work as a statistical consultant to prepare a number of studies. Lottery sales data are organized by county (there are eighty-seven counties in Loonland). Initially he wants to know whether sales revenue has grown significantly over the first six months of operation. He also wants to know which counties are the largest sellers. Are the sales per person approximately constant over the various counties, or do certain counties have higher sales per capita?

After meeting with Gerry and his staff, you conclude that your first task is to develop a data file, sorted by county, that contains the information needed to conduct your study. The lottery began operation on March 1, and you have asked for sales data for three two-month intervals: March–April, May–June, and July–August. These data indicate the number of tickets that have been sold for $1 each.

The data for this study are stored in a file named Lottery, which is stored on your data disk.[2] The names and descriptions of the variables are presented in Table 20.1. This file contains the basic data. To carry out your analysis, you may need to compute additional variables, using these basic data.

After developing the data file, you must study the growth patterns in sales over the first three periods. First you must determine whether per capita sales

---

[2]The data for this case were obtained from a former student, Mark Belles, who prepared them for his senior project. His contribution is gratefully acknowledged.

TABLE 20.1	VARIABLE NAMES IN THE LOTTERY DATA FILE	

VARIABLE NAME	NUMBER OF OBSERVATIONS	VARIABLE DESCRIPTION
County	87	County name (alpha data)
Populatn	87	Population in the county
Ctyloctn	87	Geographic location of county
Codes		
		1 = Metropolitan statistical survey area
		2 = Northwest region
		3 = Northeast region
		4 = Southeast region
		5 = Southwest region
Totalinc	87	Total income
Inc/cap	87	Per capita income
Tickets1	87	Number of tickets sold, March–April
Tck/cap1	87	Per capita tickets, March–April
Tickets2	87	Number of tickets sold, May–June
Tck/cap2	87	Per capita tickets, May–June
Tickets3	87	Number of tickets sold, July–August
Tck/cap3	87	Per capita tickets, July–August
Unemploy	87	Percentage unemployment
Highschol	87	Percentage with high school education in county
College	87	Percentage four-year college degree in county
Touriste	87	Total tourist expenditures
Age15-24	87	Percentage age 15–24
Age25-64	87	Percentage age 25–64
Age65	87	Percentage age 65 and above

have significantly increased over the first three two-month blocks of time. You must also find out whether any counties have failed to experience a growth in lottery ticket sales.

In this case you will prepare three separate descriptive analyses for the early life of the lottery: initiation (March–April), growth (May–June), and steady-state operation (July–August). A good measure of performance for the lottery business is per capita ticket sales. This measure indicates both market penetration and total sales, because the population of the state is stable. The data file identifies per capita sales for each of the eighty-seven counties in the state.

Most business operations experience some instability during their start-up period. However, successful businesses correct this problem as they move toward a steady-state operation. Thus a successful business should both grow and reduce instability during its initial operating period. Per capita sales differ among counties because of different behavior patterns by residents of these counties. Such differences are influenced by local culture, institutional norms, and the activities of the citizens.

Initially some of the variation in per capita sales is attributable to differences in business practices across various counties. For example, there may be too few retailers selling tickets, or store clerks may be inadequately trained. Alternatively, the equipment for issuing tickets, communicating with the state office, and selecting winners may experience operating problems. After some time, these problems should be corrected.

Your analysis in this case should be directed toward determining the improvements in sales stability and in sales growth during the early period of the lottery's operation.

After your series of meetings, Gerry and you prepare the following list of analysis activities.

20.1 Load the Lottery data file into your local computer system.

20.2 Analyze the growth of lottery sales over the first three bimonthly intervals.
  a. Prepare histograms of per capita sales for each of the three time periods. Discuss the differences among sales patterns for the three intervals.
  b. Prepare a scatter plot of per capita sales for the first period and the third period. Could third period per capita sales be predicted from first period per capita sales? What does the graph tell you about the growth in sales from the first period to the third period?
  c. Perform a statistical hypothesis test to determine whether statistically significant per capita sales growth has occurred from the first to the third period.

20.3 Using July and August sales data, identify the counties in the upper quartile of total sales, and determine their contribution to total state sales.
  a. List the counties in the upper quartile by name, along with total sales, per capita sales, and population.
  b. What percentage of total state sales comes from these counties?
  c. What percentage of state population is in these counties?
  d. Compare per capita sales in these counties with per capita sales for the entire state.

20.4 Prepare a written report for Gerald O'Leary that discusses your analysis and presents a description of the lottery during its early operating period. Indicate your conclusion regarding progress toward sales growth and reduced variability during this period.

CASE 21

STATE LOTTERY B

Variables Influencing Lottery Sales

21.1 BACKGROUND

In recent years, voters have expressed a strong desire for lower taxes—without, however, expressing a commensurate "demand" for reduced or eliminated services. In partial response to this problem, a number of state legislatures have passed laws to establish a state lottery. The rationale is simple: people like to gamble; legal state-run gambling can be controlled to avoid negative side effects; and the state should take advantage of this situation to raise revenue without having to raise taxes.[1]

The State of Loonland recently established a state lottery and has hired an experienced lottery executive, Gerald O'Leary, to install and operate it. As part of the development plan, Gerry set up a data collection system to record sales by county. This system permits him to monitor the lottery operation from start-up onward. The lottery operation consists of selling, at various retail outlets, game cards that offer the possibility of winning an instant prize, with a specified probability. In addition, card numbers are used for a random selection of winners of much larger daily and weekly prizes. If no winner is identified for a daily or weekly prize, the prize pool grows, providing an opportunity for publicity to increase excitement and sales among the pool of prospective players. The probabilities associated with winning various prizes are established so that the expected return to players is 50 percent of total sales revenue. Administration of the state lottery apparatus consumes 10 percent of revenue, and the retail outlets receive 5 percent. Thus, 35 percent of total sales revenue is delivered to the treasury of Loonland. In the original bill establishing the lottery, this money was earmarked for conservation, pollution abatement, and education. In addition, a small amount—approximately 0.5 percent of revenue—has been designated for treatment of compulsive gamblers.

---

[1]Many states have developed state-run lotteries, so the legitimacy of this activity is not at issue here. Our concern is with the use of statistical analysis to assist in the managing of a state-run business. This case is based on real data, but all proper names are fictitious.

Gerald O'Leary has established a staff of strong professionals to develop and administer the new lottery. As time goes on, the lottery staff will continually devise new lottery game formats and will prepare advertising and promotional material to maintain a high level of interest among the pool of regular players. The actual game cards and the procedures for assigning the probability of winning are purchased from a supplier who guarantees the appropriate probabilities for the various prizes. Staff members also negotiate with retailers who apply to be sales outlets. The lottery staff trains store personnel to use equipment that issues the cards in each store. After a retailer has signed a contract as an authorized lottery sales outlet, the lottery staff regularly monitors each outlet to ensure good relations with players and correct reporting of revenue. Gerry has made his staff aware of the importance of good relationships between players and retailers. In addition, lottery administrators have developed procedures for monitoring sales to ensure proper reporting of sales revenue from each retail outlet.

21.2 MANAGEMENT ISSUES AND OPERATING DATA

Gerald O'Leary knows the importance of good statistical analysis for monitoring the lottery start-up and its continuing operation. He has asked you to work as a statistical consultant to prepare a number of studies. Lottery sales data are organized by the eighty-seven counties in the state of Loonland. Gerald wants to know whether there is any relationship between lottery sales and tourism expenditures. Essentially, should the lottery commission develop sales promotions and possibly even special games that are directed toward tourists? In addition Gerry wants to know whether per capita sales are related to income, education, or unemployment levels in the counties.

After meeting with Gerry and his staff, you conclude that your first task is to develop an analysis data file, sorted by county, that contains the information needed to conduct your study. The lottery began operation on March 1, and you have asked for sales data for three two-month intervals: March–April, May–June, and July–August. These data indicate the number of tickets sold for $1 each. To complete the analysis, you have asked the staff to collect additional data by county on tourism expenditures, income, and education. These data are stored in a file named Lottery, which is stored on your data disk.[2] The names and descriptions of the variables are presented in Table 21.1. This file contains the basic data. To carry out the analysis, you may need to compute additional variables, using the basic data.

---

[2]The data for this case were obtained from a former student Mark Belles, who prepared them for his senior project. His contribution is gratefully acknowledged.

TABLE 21.1		VARIABLE NAMES IN THE LOTTERY DATA FILE
VARIABLE NAME	**NUMBER OF OBSERVATIONS**	**VARIABLE DESCRIPTION**
County	87	County name (alpha data)
Populatn	87	Population in the county
Ctyloctn	87	Geographic location of county
Codes		
		1 = Metropolitan statistical survey area
		2 = Northwest region
		3 = Northeast region
		4 = Southeast region
		5 = Southwest region
Totalinc	87	Total income
Inc/cap	87	Per capita income
Tickets1	87	Number of tickets sold, March–April
Tck/cap1	87	Per capita tickets, March–April
Tickets2	87	Number of tickets sold, May–June
Tck/cap2	87	Per capita tickets, May–June
Tickets3	87	Number of tickets sold, July–August
Tck/cap3	87	Per capita tickets, July–August
Unemploy	87	Percentage unemployment
Highschol	87	Percentage with high school education in county
College	87	Percentage four-year college degree in county
Touriste	87	Total tourist expenditures
Age15-24	87	Percentage age 15–24
Age25-64	87	Percentage age 25–64
Age65	87	Percentage age 65 and above

Gerald O'Leary believes that people on vacation are more likely than other people to purchase lottery tickets. Thus he would like to target special promotional activities toward tourists. However, this strategy will be developed only if evidence indicates that tourists are more likely to purchase lottery tickets. You agree that July and August lottery sales in high-tourism areas are the ones most likely to be related to tourist activity. Thus, for example, you could determine whether increased sales in July and August above May and June sales in certain counties are related to higher tourism expendiures.

The lottery commission and lottery management also want to know whether counties with higher income and/or higher education levels tend to have higher per capita sales of lottery tickets. The answers to these questions could be used to develop advertising and promotion material with a more specific target. In addition Gerry wants to counter the claim made by some critics that the lottery is an "unfair tax" that tends to draw a disproportionate amount of money from persons with lower income and/or education. Currently such complaints are minimal; but negative campaigns could mount quickly so Gerry wants to be prepared with a clear understanding of the facts.

You decide that these questions can be handled best by developing a multiple regression model that predicts per capita lottery sales by county, using per capita income, a measure of education level, and other suitable variables. The data contain two measures of education; the percentage with high-school educations or higher, and the percentage with four-year college degrees or higher. Since no clear theory supports either of the education measures, you decide to try them both and to use statistical procedures to determine which provides the better prediction model when combined with income.

After your series of meetings, Gerry and you prepare the following list of analysis activities.

21.1 Load the data from the Lottery file into your local computer system.

21.2 Analyze the data to determine whether tourism has an effect on lottery ticket sales. Use appropriate statistical tests and prepare a one-page written report describing your analysis and presenting its conclusions.

21.3 Perform an analysis to determine whether per capita lottery ticket sales differ between urban and rural populations. Prepare a short report describing your analysis and presenting the conclusions.

21.4 Perform an analysis to determine whether per capita lottery ticket sales differ between urban population and each of the four geographic regions of the state. Prepare a short report discussing your analysis and indicating its conclusions.

21.5 Determine whether a relationship exists between counties with high per capita lottery sales and the average income and educational levels in the county.

   a. Prepare scatter plots of per capita lottery sales for the May–June and July–August periods versus income levels and education levels. Try both the percentage of high school graduates and the percentage of college graduates as the measure of education, and use the measure that proves to be the stronger predictor in the remaining analysis.

   b. Develop a model that predicts per capita lottery sales as a function of income and of educational level. Present the model results in a standard form, and explain your conclusions.

   c. Use the model developed above to determine whether significant differences exist in per capita lottery sales among the major regions of the state after you have adjusted for income and education.

   d. Beginning with the model developed thus far, determine whether other variables in the data file are conditional predictors of per capita lottery sales.

21.6 Prepare a written report for Gerry discussing your analysis. Advise Gerry about the possible relationship between tourism and lottery purchase. Discuss how he should respond to potential critics of the lottery based on the results of your analysis. Indicate the facts clearly, and indicate to him the strength of the results.

NATIONAL CRIME CONTROL PROGRAM

Crime Rate Changes

22.1 BACKGROUND

Concern has grown in the United States about the high levels of crime. Urban dwellers often do not leave their homes at night for fear of being subjected to violence. Loss of property concerns both rural and urban people. Political candidates promise that they will get tough on criminals and reduce crime rates.

Crime rates are the result of a complex process of human behavior whose complete analysis and solution is beyond the scope of this statistical study. However, the problem addressed here is a typical one of public-sector policy, which attempts to deal with complex social issues. In response to public pressure, the federal government established a specific set of remedial programs, based on research papers and public testimony prepared by persons who have studied the problem for a number of years. In spite of careful planning, the success of the program is uncertain. Thus an important part of program development is an evaluation to determine whether the programs have achieved their objective. In this case you will evaluate a program designed to reduce crime rates.[1]

In response to a campaign pledge, the President proposed and Congress passed a national crime reduction bill. The Justice Department was directed to develop a program to assist local law enforcement agencies in their fight against crime. The program provided a substantial fund to support local government matching grants. Local police developed projects and applied for matching fund grants from the Justice Department fund. Eligible projects could include additional police officers, youth recreation and education programs, special police equipment, and improved statewide programs to

---

[1]This case uses real data and a standard analysis format. The actual data were obtained for the years 1982 and 1984. Assume that your analysis will apply to programs during those years. The case is not designed to provide support for or opposition to any specific program; instead, it is designed to provide a realistic evaluation that includes problems with assumptions, data, and the choice of specific statistical analysis procedures.

coordinate information about criminal activity. In addition, the federal program developed a national computer-accessible criminal data bank that included finger print identification of persons with records of criminal activity. Local police can routinely check all arrested persons for other criminal records.

The national crime reduction program has been criticized by a number of House and Senate members. They argue that the program is very expensive but has not reduced crime rates, and thus is a waste of tax revenue. As a result of these criticisms, the Department of Justice has commissioned a study to determine whether crime rates have been reduced. To continue its support for the program, the Department of Justice needs strong evidence that crime rates have been reduced. If this evidence cannot be obtained, the Justice Department has indicated that the program will be canceled.

22.2   ANALYSIS OF CHANGES IN CRIME RATES

In this case you are to design and implement an evaluation of the national crime reduction program. You will design an appropriate hypothesis test and use actual data to conduct the analysis. Your analysis will apply appropriate statistical procedures and also consider potential problems with data.

A major portion of this program was implemented in 1983. Your task is to determine whether national crime rates changed between 1982 and 1984. You are thus conducting a before and after study, using observations of crime rates from each of the fifty states and the District of Columbia. If crime rates decreased and other major changes did not occur at the same time, one might reasonably infer that the program was a success.[2] Assume that other studies have investigated the occurrence of other events that might have been related to the change. Based on those studies, you conclude that other events did not have a major influence on crime rates over the period from 1982 through 1984.

To conduct this evaluation, you must develop an analysis strategy. This strategy includes defining variables and selecting the population to be represented by the study. These analysis design decisions are influenced by the study objectives, available data, the state of knowledge concerning the problem, and the creativity of the research team. Design activity usually accounts for a major portion of actual research projects. Typically it begins with an extensive review of the literature and discussions with persons who have studied the problem. In many cases the preliminary analysis generates working papers and documents. These papers are discussed and revised extensively by the research team and by outside reviewers and consultants. In

---

[2]For example, crime rates are usually higher among the portion of the population aged 16–25. Thus, if the population proportion from this age group decreases the overall crime rates will decrease.

addition, these documents are reviewed and revised at various times during the study. Most research involves cycling between the study design, data collection, analysis, and writing stages. In this project, the study design and data collection have been completed. You will concentrate on the analysis and writing stages of the study.

Beginning with the initial concern about crime, the research team has made some initial decisions. The study is to represent the entire United States, and crime rates for individual states will be the observations. State data provide a representation of a wide range of societal characteristics and geographic regions. Because state populations vary considerably, crime rates per 100,000 people are used to provide comparable measures for each state.

This study will assume that the crime rates in the sample have the characteristics of a random sample, thus enabling us to perform statistical tests that are based on random samples. The decision to treat the sample of data as a random sample is based on the following rationale. Given the process of collecting crime data, the reported crime rate from each state clearly contains a sizable error component that has random properties. Thus if the data collection process were repeated, a different set of crime rates for each state would probably be reported. The study must consider the inherent unexplained error component in state crime rates.

The distribution of state crime rates comes from a symmetric distribution such that sample means follow a normal distribution, based on the Central Limit theorem. These assumptions, which are based on the experience of statisticians who have worked with these data, enable us to apply the tools of hypothesis testing to this problem. (These assumptions would not be accepted by all statisticians.)

Data on state crime rates for a particular time period will be analyzed, and the results will be applied to the long-term problem of crime rate reduction. Thus this study can be classified as an analytical study. The data are assumed to come from implicit populations of past and future crime rates.

22.3 ANALYSIS PROCEDURE

In your study you will compare all of the 1982 and 1984 crime rates. This will require you to use hypothesis tests to determine whether the rates have increased, decreased, or remained unchanged.

The data were obtained from the 1984 and 1986 Statistical Abstracts and stored in a data file named Crime. This data file contains 1982 and 1984 crime rate data for all fifty individual states and the District of Columbia. The crime rates are expressed in reported crimes per 100,000 population, as collected by the Federal Bureau of Investigation. Variables in the file are described in Table 22.1. Examine this description to determine the definition of the variables.

TABLE 22.1	VARIABLE NAMES IN THE CRIME DATA FILE

VARIABLE NAME	NUMBER OF OBSERVATONS	VARIABLE DESCRIPTION
Code	51	State code
Totcri82	51	Total crime rate 1982
Totcri84	51	Total crime rate 1984
Murd82	51	Murder rate 1982
Murd84	51	Murder rate 1984
Rape82	51	Rape rate 1982
Rape84	51	Rape rate 1984
Asslt82	51	Assault rate 1982
Asslt84	51	Assault rate 1984
Rober82	51	Robbery rate 1982
Rober84	51	Robbery rate 1984
Burgl82	51	Burglary rate 1982
Burgl84	51	Burglary rate 1984
Larcny82	51	Larceny rate 1982
Larcny84	51	Larceny rate 1984
Mrtveh82	51	Motor vehicle theft rate 1982
Mrtveh84	51	Motor vehicle theft rate 1984

Notice that a code identifying each state is included in the list of variables. This identification permits you to trace specific data back to a state if you discover some unusual results during the data analysis. A listing of the state codes and state names is provided in Exhibit 22.1.

The data for this study were obtained from state reporting sources, and thus there is the possibility of systematic errors in recording data. These are usually recognized by their excessive deviation from the typical values of the data. Careful researchers examine the data for unusual observations. Typically, such observations are either removed or adjusted, based on other information. In this study you are expected to use your judgment, based on what you know about crime rates in different states. You may also ask faculty and others who are likely to have some understanding of reported crime rates. The specific crime rates for this study include total crimes per 100,000 people, murder, rape, assault, robbery, burglary, larceny, and motor vehicle theft.

Your specific tasks in developing an analysis of the 1982 and 1984 crime data include the following steps.

22.1 Load the data from the Crime file into your local computer system.

22.2 Assuming that the 1982 and 1984 crime rates are independent, use an appropriate hypothesis test to determine whether rates of total crime and of disaggregated crime have decreased between 1982 and 1984. You should conclude that crime rates have decreased only if the probability of making an error in reaching this conclusion is 0.05 or less.

EXHIBIT 22.1	STATE CODES	

1. Maine	18. North Dakota	35. Arkansas
2. New Hampshire	19. South Dakota	36. Louisiana
3. Vermont	20. Nebraska	37. Oklahoma
4. Massachusetts	21. Kansas	38. Texas
5. Rhode Island	22. Delaware	39. Montana
6. Connecticut	23. Maryland	40. Idaho
7. New York	24. Dist. of Columbia	41. Wyoming
8. New Jersey	25. Virginia	42. Colorado
9. Pennsylvania	26. West Virginia	43. New Mexico
10. Ohio	27. North Carolina	44. Arizona
11. Indiana	28. South Carolina	45. Utah
12. Illinois	29. Georgia	46. Nevada
13. Michigan	30. Florida	47. Washington
14. Wisconsin	31. Kentucky	48. Oregon
15. Minnesota	32. Tennessee	49. California
16. Iowa	33. Alabama	50. Alaska
17. Missouri	34. Mississippi	51. Hawaii

Otherwise, you should conclude that insufficient evidence exits to permit you to reject the hypothesis that there has either been no change or an increase in crime rates. Formulation of the correct hypothesis test requires careful thinking and definition of the requirements. Present your conclusions using appropriate hypothesis tests, which clearly indicate your analysis.

22.3 In question 22.2 you assumed that the observations in the two samples were independent. Because both samples use states as the observations, however, correlated observations are possible. This would occur, for example, if states with low crime rates in 1982 tended to have low rates in 1984; and vice versa for high-crime-rate states. Since variances of differences between sample means are smaller for correlated samples, it is important to determine whether the samples are correlated.

a. Plot the 1982 versus 1984 crime rates for all crime categories. By examining the plots, indicate your initial conclusion.

b. Compute correlations between the samples for each crime rate, and perform a hypothesis test to determine whether the crime rates are correlated

22.4 Repeat the hypothesis tests from question 22.2 using the assumption either of independent or of correlated samples, depending on your analysis in question 22.3. Present your new results clearly, and discuss any differences.

22.5 Prepare a report that presents the results of your research. Include specific recommendations regarding the continuation of the crime control program, based on your statistical analysis. The total written report should not exceed two pages. Include carefully prepared exhibits and statistical appendices to support your analysis. Finally, prepare a half-page executive summary of your study, to appear at the beginning of your report, concisely presenting your conclusions and briefly (3 or 4 sentences) describing your approach and method. The reader should understand your results after reading the summary.

CASE 23

AMERICAN MOTORS INC.

Predicting Fuel Economy and Price

23.1 INTRODUCTION AND BACKGROUND

American Motors Inc. is a major national manufacturer of automobiles and light trucks. It competes with one other large domestic company and two smaller firms. Significant competition also comes from European and Japanese companies. The industry has undergone major changes, including an emphasis on fuel economy mandated by both national government standards and market demand from an increasing number of consumers. Engineering and production staff have developed numerous improvements in response to market demands. Marketing research has continued to expand the company's understanding of customer needs and desires.

Foreign competition resulted in major sales reductions for all domestic U.S. manufacturers. When it experienced those reductions, American Motors was forced to close a number of facilities and reduce employment. In addition, new design and production innovations had to be implemented to obtain the same cost and quality parameters as those of the foreign manufacturers.

American Motors has committed considerable resources to new facilities for product development, quality control, and manufacturing. As a result, quality has improved and productivity has increased. However, the costs of these improvements has added large debt service charges to annual costs. Thus sales levels must be maintained and increased if possible.

23.2 LONG-RANGE PLANNING ANALYSIS

After years of declining sales and facility downsizing American Motors has begun to experience increased demand for its vehicles. Design, performance, and quality measures indicate that its vehicles compare favorably with those produced by other foreign and domestic companies.

Reflecting on the years of major business problems, senior management realized that during the 1970s it had not been sensitive to the business environment. Consumer tastes changed, and foreign competitors produced cars that responded better to consumer needs. In addition American Motor's quality had decreased so that performance was substandard for the industry. To prevent future problems, American developed a long-range market planning group responsible for monitoring the market environment and recommending change to help maintain American's competitive position.

The long-range market planning group prepares alternative strategies for developing future cars, and it monitors American's performance compared to the rest of the industry. Each company in the industry has various automobile models that resemble those in American's product mix. However, the number of units sold for each of various models differs for each company. Therefore, it is difficult to compare companies by the product mix. Instead the market planning group has developed comparisons based on important vehicle performance characteristics, including number of cylinders, horsepower, acceleration, engine displacement, and vehicle weight.

You have been asked to develop mathematical models to predict fuel economy and vehicle selling price as a function of these performance characteristics. These models will be used to estimate which performance characteristics have the greatest effect on purchasing decisions. In addition, the effect of various combinations of characteristics will be estimated. A representative sample of the various automobiles in the present national vehicle mix has been obtained for your analysis.

The analysis will determine which performance variables are significant predictors and will estimate the linear coefficients of those variables. Before beginning the analysis, you meet with a number of experienced engineers to learn more about the vehicle performance characteristics and their relationship to fuel economy. You also talk with a number of experienced members of the marketing staff to learn how various consumer groups rate the importance of the different performance characteristics.

The fuel economy variable has become increasingly important for vehicle marketing. Initially, fuel economy represented an important national policy objective after major producer countries restricted supply and increased price. The national policy contained a number of energy conservation measures, including fuel economy (miles per gallon) minimums, imposed on each manufacturer. A manufacturer can meet these standards either by selling a greater proportion of small fuel-efficient cars or by improving the fuel economy of its larger and higher-priced cars. The company would prefer the latter strategy, because larger cars provide higher revenue and a larger contribution to overhead and profit. Fuel economy improvements can be obtained by reducing vehicle size and weight, by reducing engine size and horsepower, or by providing a combination of weight and horsepower reductions. Having smaller size and weight usually reduces comfort, while the lower horsepower reduces vehicle performance. Fuel economy can also be

TABLE 23.1	VARIABLE NAMES IN THE MOTORS DATA FILE	

VARIABLE NAME	NUMBER OF OBSERVATIONS	VARIABLE DESCRIPTION
Observtn	155	Code identifying car observations
Milpgal	154	Miles per gallon
Cylinder	155	Number of cylinders in the car
Displace	155	Cubic inches of engine displacement
Horspwr	151	Horsepower generated by the engine
Accel	155	Acceleration (in seconds) to 60 miles per hour
Year	155	Model year of the vehicle
Weight	155	Vehicle weight (in pounds)
Country	155	Country of origin (1 = U.S., 2 = Europe, 3 = Japan)
Price	155	Price in dollars
Company	155	Manufacturer (1 = American Motors; 2 = Apex Motors; 3 = Smith Motors; 4 = Other)

improved by instituting better engine design and engine operating control. For example, the combination of sensors and a computer processor control can provide the ideal fuel/air mixture for different vehicle load conditions. The present overall vehicle mix results from a number of consumer choices. Thus, analysis of the present mix offers a way of measuring consumer preferences.

After meeting to define specific objectives for the study, the planning staff decides that the study should identify key factors that predict fuel economy and key factors that predict vehicle price. These factors could then be used as planning parameters for developing new vehicle designs. American Motors also wants to know how it compares with the rest of the industry as gauged by the predictive factors, and it wants to know the relationship of these factors to fuel economy and vehicle price.

The planning group has asked you to perform an analysis to answer these questions. The data file for these variables is described in Table 23.1, and the data are contained in the file named Motors. Your analysis should include the following steps.

23.1 Load the data from the Motors file into your local computer system.

23.2 Identify which factors significantly predict fuel economy for the entire sample.

23.3 Determine whether the fuel economy for American Motors vehicles is significantly above or below—or roughly matches—the industry level, with appropriate adjustments for differences in the mix of the predictive factors.

23.4 Indicate what factors significantly predict vehicle selling price for the entire sample.

23.5 Determine whether the selling price for American Motors vehicles is significantly above or below—or roughly matches—the industry level, with appropriate adjustments for differences in the mix of the predictive factors.

CASE 24

STATE PLANNING FOR LOONLAND

Small City Economic Development

24.1 INTRODUCTION

You have been asked to perform data analysis and statistical model development for the state planning agency in the State of Loonland. The state legislature has requested a major study of economic development in small cities. Rebecca Glover, director of state planning, has developed a design for the entire study that includes statistical analysis to determine how economic development influences tax rates and property values in small cities.

As a first step Rebecca called a meeting of small-city leaders, from across the state and asked them to talk about their concerns and solutions. From that discussion, a basis for developing study questions emerged.

The small city leaders were concerned about the economic strength and growth potential of their communities. Some small cities had stagnant economies, although the state as a whole has experienced significant growth (most of it near the state's main metropolitan area). Commercial and industrial growth has expanded greatly across all segments of the metro region, and many younger people have moved there to obtain employment. An increasing number of small-city residents drive to the urban area for retail purchases and entertainment, believing that the number of options in the urban area is much greater than at home. Because so many dollars are leaving these communities, some small-city stores have closed. Small-city mayors were very concerned about property tax rates and the difficulty of obtaining money for local government services. They argued that economic expansion is essential to provide a larger property base on which taxes can be levied.

However, it was also noted that small cities tend to have a slower-paced life-style that many residents appreciate. Traffic congestion and crime rates tend to be less of a problem. Many argue that the small city is a better place for families and that children can grow up in a more secure and relaxed environment. Thus some people—especially those with good jobs—were quite happy with their small city. They would prefer limited growth so that their communities can avoid acquiring the problems of the larger metropolitan region.

24.2 DEVELOPMENT OF ANALYSIS

From your initial discussions with various people, you have begun to develop a number of questions for your study. Total tax revenue for local government services is the product of the tax rate times the total assessed property value. Increased revenues require either an expansion of property value or an increase in the tax rate. The latter implies that each resident will pay higher taxes. In recent years, the total tax paid by individuals has become a volatile political issue; and many political campaigns have been based on promises to reduce the property tax rate. Thus local elected officials generally prefer to expand the property base rather than to raise tax rates.

The property base can be expanded by increasing the number of homes, apartments, commercial establishments, or industrial property. The former two result in more residents, although they also increase demand for city services (and hence for more revenue). The latter two create employment and thus could lead to more residents who require more houses and apartments. In addition, business property—especially commercial property—creates its own demand for additional government services, such as police and fire, roads, city planning, and zoning. The problem becomes more complicated because one needs to know whether certain kinds of property foster greater demands for local government services. For example, does a new shopping center require more police and an expanded road system? Do store owners demand more government services than factory owners and/or home owners? Do apartment residents use more city services than homeowners? When considering various expansion options, you must determine how different types of property influence the local tax rate.

Economic expansion influences the quality of life in small cities. Certain kinds of expansion—for example, a dirty high-pollution factory—are likely to reduce quality of life. In contrast, a nice restaurant or an industry that provides attractive jobs is likely to increase quality of life. Quality of life is difficult to measure, but the quality of a neighborhood or community generally influences the price of housing of a given size and style. Many people will pay a higher price and/or occupy a smaller house in order to live in a more desirable community. Therefore, market value of houses, adjusted for size, could be a measure of quality of life in the community. Economic expansion that leads to increased housing values improves quality of life in the community.

As a first step in your study, you have collected data for individual cities from the state finance office. The data represent 45 small cities outside the metropolitan region, and they cover two different years, thus providing a total of 90 observations.[1] The data are contained in a file named State, and the

---

[1]These data were actually collected as part of a research project conducted by the author during the 1970s. The government policy questions discussed remain relevant, and the data represent the actual situation at that time.

TABLE 24.1	VARIABLE NAMES IN THE STATE DATA FILE

VARIABLE NAME	NUMBER OF OBSERVATIONS	VARIABLE DESCRIPTION
Obsnum	90	Sequential observation number
County	90	Code for the county
City	90	Code for the city
Sizehse	90	Median rooms per owner-occupied house
Totexp	90	Total current city government expenditures
Taxbase	90	Assessment base (in millions of 1972 dollars)
Taxrate	90	Tax levy divided by total assessment
Pop	90	City population estimate
Income	90	Per capita income (in 1972 dollars)
Hseval	90	Average market value per owner-occupied residence
Taxhse	90	Average tax per owner-occupied residence
Homper	90	Percent property value: owner-occupied residence
Rentper	90	Percent property value: rental residence
Comper	90	Percent property value: commercial
Indper	90	Percent property value: industrial
Utilper	90	Percent property value: public utility
Year	90	Year represented by the data

variables are described in Table 24.1. Included in the file are measures of total property in the community by major classification, tax rate, housing price, housing size, and personal income.

After your preliminary investigation and discussion, you meet with Rebecca Glover and her staff to develop study objectives. Based on these meetings and the legislative mandate, you decide that the study should develop recommendations for small-city growth. These recommendations will indicate the type of development that should be encouraged, consistent with the joint goals of lower tax rates and higher quality of life, as measured by housing prices. Your report should indicate the effect of different kinds of development on the two outcome goals and should discuss the alternatives that emphasize one or the other goal.

Your study should generally proceed through the following steps.

24.1 Load the State data file into your local computer system.

24.2 Prepare descriptive statistics for the analysis variables, including a correlation matrix. Prepare a short discussion that describes the variables and indicates any unusual patterns.

24.3 Develop a model to predict tax rate as a function of other variables related to economic development policy.

24.4 Develop a model to predict market value of owner-occupied houses as a function of variables related to economic development policy.

24.5 Compare the two final models and indicate which recommendations would be similar and which ones require a trade-off between goals.

24.6 Using your models, study the patterns of residuals and their relationships to other variables. Identify any special relationships or conditions that might provide valuable recommendations for local or state planning policy.

24.7 Prepare a report discussing your results and providing recommendations to local communities regarding their economic development policies. Provide a clear rationale for your recommendations, based on your statistical results, and make the recommendations comprehensible to local government officials who do not have your level of statistical skill.

CASE 25

PRODUCTION SYSTEMS INC.

Development of a Salary Model

25.1 INTRODUCTION

Sally Parsons, president of Production Systems Inc., has asked you to assist in analyzing the company's salary data. She has recently received a series of complaints that women employees are receiving lower wages for comparable jobs. The complaints came as a surprise to Sally because she thought salary increases were based on experience and performance. Sally was aware that the average wages of women staff were less overall than those of men. However, she also knew that women staff had less experience and thus would be expected to have lower wages. Given the complaints Sally understands that she must have objective information.

Production Systems Inc. is a regional computer systems development company that specializes in work with banks and insurance companies. The company started as a service department for a regional accounting company in the 1960s. A few of the current employees actually came from the accounting company. In the early 1970s, Production Systems Inc. became an independent company. The company has been quite successful, experiencing steady growth over its entire life. Employment has not grown as rapidly as have the company's total billings because of productivity innovations introduced during the past ten years.

The company has tended to hire experienced professionals with masters degrees in technical fields including business, management science, computer science, engineering, economics, and mathematics. Most of the employees have come with significant experience. The youngest employee in the professional group is 29, and the most senior is 65. Experience with Production Systems varies over a wide range, with one-fourth having less than seven years and one-fourth having more than 22 years. Most of the women employees have less experience with the company.

The professional staff has only three levels: systems analyst, team systems analyst, and project systems analyst. Salary ranges are quite wide within each of the levels. Project systems analyst is the highest level, followed by team systems analyst. Promotion to the higher ranks is awarded by a committee of project systems analysts, and advancement to each level usually requires a minimum of six years' experience and significant project work. In general, persons at the higher levels are more productive and tend to direct projects. It is possible, however, for a group that includes several project systems analysts to be directed by a team systems analyst. For every project under contract, a team of the best available people is created to carry out the work. It is also well known that some persons at the highest level are less productive than others at lower levels. Thus higher status and salary is a reward for past performance and not a reliable measure of present contribution.

Salary adjustments are sometimes made to recognize certain specialty skills that demand a high price in the labor market. Persons who work in database systems programming have unique skills that are highly sought by other companies. Another special category is technical systems developers—people who prepare specialized high-performance software for key parts of large systems. People with either of these skills are in great demand, so they must be paid a premium if they are to be retained. Such specialists work at all three professional staff levels, depending on their experience with the company, but the company has not provided premium salaries merely by promoting the specialists. The personnel policy has been to base promotions on a wide range of work and project management skills. Special skills are compensated by a separate adjustment. Because promotions to higher levels are related to experience, the company has sought to avoid confusing levels and specialized skills that have a market premium.

25.2 PROBLEM ANALYSIS

Your project analysis begins with a series of meeting you have with Sally Parsons and the director of human resources, Gilbert Chatfield. Both Sally and Gilbert indicate that wages tend to increase with experience in the company. The managers conduct an annual employee review, which relies heavily on input from project leaders who are directing teams at various remote locations. Project leaders shift as projects are completed and new teams are assigned. Thus, obtaining consistent information to provide the basis for a high-quality employee evaluation is difficult. Most of the employees at Production Systems enjoy their independence and challenging work; salary levels have not been a major concern for most employees. Certain people are recognized as strong performers, and their increases and promotions are generally accepted by the professional staff.

In recent years, however, concerns have been raised about the fairness of the system of awarding salary increases. The complaint by women employees is the most serious, but other complaints have been made over the past

TABLE 25.1	VARIABLE NAMES IN THE PRODSYS DATA FILE		

VARIABLE NUMBER	VARIABLE NAME	NUMBER OF OBSERVATIONS	VARIABLE DESCRIPTIONS
1	Age	150	Age of the employee
2	Yearsexp	150	Years of experience with Production Systems
3	Yearslv2	150	Number of years as a team systems analyst
4	Yearslv3	150	Number of years as a project systems analyst
5	Female	150	Gender: 1 = female; 0 = male
6	Salary	150	Annual salary (in dollars)
7	Speclty1	150	Specialty: 1 = Database systems development skill; 0 = else
8	Speclty2	150	Specialty: 1 = Technical systems development skill; 0 = else

several years. In view of these concerns, you recommend that a salary prediction model be developed. This model would use data based on the current salaries paid to professional staff and important variables that define the experience and skill levels of the staff. Such a model would indicate the effect of various factors that contribute to salary level, and it would identify persons whose salaries are above and below the predicted average salary. The model could also be used to determine whether an employee's gender predicts a salary that is higher or lower than would be expected on the basis of experience and qualifications.

After some discussion, Sally and Gilbert agree that this model should be developed. It would be useful for answering the present complaint, and it would provide a tool for reviewing the complete professional staff salary structure. After reviewing the employee data records, you select a candidate set of variables for the model development. These variables, which are contained in a file named Prodsys, are described in Table 25.1. To protect the confidentiality of each employee's salary record, there is no variable to identify individual employees in this file. At the completion of the study, you will provide Gilbert with a list of employees who are substantially below or above the standard predicted by the model. Since he has the identification key for each employee and has access to other performance information, he can decide whether certain persons' salaries should actually be above or below the standard.

Your final discussion concludes with your agreeing to include the following tasks in your study.

25.1 Load the data from the Prodsys file into your local computer system.

25.2 Prepare descriptive statistics and graphs for the variables in the data file. Is the average compensation paid to all women less than that paid to all men?

25.3 Consult textbooks and other references to help specify the variables and functional form for the salary prediction model. Provide a short discussion of this literature review.

25.4 Use multiple regression analysis to develop a salary prediction model for Production System's professional staff.

25.5 Use the preceding model to determine whether gender discrimination exists in the salary structure.

25.6 Prepare a written report presenting the salary prediction model and indicating its strengths and weaknesses. Then discuss the wage discrimination complaint and your conclusion regarding its validity. Include appropriate summary data and graphs to help communicate your conclusion.

SHELDAHL INC.

Experimental Design: Thickness Coating
of Entek-Plus on ASI Coater

26.1 INTRODUCTION

Sheldahl Inc., of Northfield, Minnesota, is a manufacturer of thin film materials. Historically its products have been used in various defense materials and in the United States space program. Original applications included weather balloons, high-strength tapes, insulation, and electronic shields, among others. In the space program, Sheldahl thin films were used in a number of applications to cover materials. Most of the satellite missions have included products from Sheldahl. The products also found great success in various defense applications, including a number that were classified. In recent years, reductions in defense contracts have forced the company to seek new markets for its technology.

Changes in Sheldahl's business orientation away from defense applications have utilized the company's leadership in the core technology of thin films to develop various new applications. One important developmental area involves flexible circuits, which are widely used in various computer applications and in the auto industry. Sheldahl engineers have used their basic skills to laminate thin copper to flexible plastic resin films. The resulting material is then used to produce flexible circuits for electronic applications. Circuit designs are prepared on computer-assisted design (CAD) systems. The resulting designs are photo-etched on the copper surfaces, and the excess copper is removed by immersion in a series of acidic baths.

The electronics industry is highly competitive, and supplier firms such as Sheldahl must constantly improve their product innovation, productivity, and quality control. To achieve these objectives, they use a variety of statistical procedures to develop the most productive manufacturing processes possible and to monitor the production processes. This case presents an example of the application of statistical procedures to improve productivity at Sheldahl Inc.[1]

---

[1]J. B. Munson, "Screening Experiment on Thickness of Entek-Plus Cu106A, on ASI Coater," Report No. M-69, Sheldahl Inc., Northfield, MN, November 1993.

FIGURE 26.1	SCHEMATIC DIAGRAM OF ASI COATING LINE

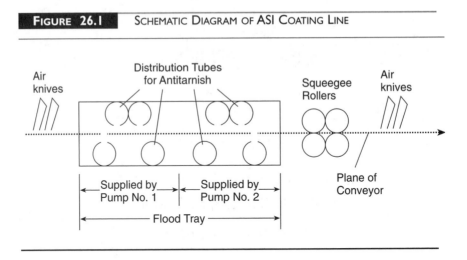

26.2 PROBLEM ANALYSIS

J. B. Munson, R & D engineer, has asked for your assistance in analyzing data from an experiment to determine the variables that affect the thickness of an organic anti-tarnish coating on copper. The coating is used to prevent oxidation on the copper surface during shipping, so that components and connections can be soldered to the circuit. Oxidation results in poor solder connections and a failure of the device that uses the flexible circuit.

The organic anti-tarnish coating is applied to continuous rolls of the copper and resin laminate on an ASI horizontal coating line, as shown in Figure 26.1. The material moves through a horizontal tray, and the coating fluid is flooded on the continuous sheet. Prior to the coating operation the sheet passes through a rinse solution to remove any oxidation and to provide a clean surface for the coating. Between the cleaning and the coating processes, "air knives" blow the cleaning solution off the copper laminate surface. Within the coating tray, two pumps flood the surface with the coating solution. At the end of the tray, soft squeegee rollers remove excess solution from the surface, followed by "air knives" that blow off any remaining solution. This coating station is followed by a final rinse in a cold water spray located downstream from the coating station.

The objective of the experiment was to identify variables that influence the thickness of the coating material and to ascertain the amount of their influence. Based on previous knowledge of the process, J. B. indicated that the experimental design would test for linear effects of the various variables. Thus, in the experimental design, each variable was measured at only two levels in combinations that would make it possible to test for direct effects using multiple regression with dummy variables. Staff members from pro-

TABLE 26.1	VARIABLES USED IN THE EXPERIMENTAL DESIGN			
NAME	VARIABLE DESCRIPTION	UNITS	LOW VALUE	HIGH VALUE
Webspeed	Web (Conveyor) speed	inches/min.	10	60
Bathtempt	Flood module bath temperature	degrees. F	100	120
P1speed	Speed of flood module pump no. 1	% full speed	5	50
P2speed	Speed of flood module pump no. 2	% full speed	5	50
Orient	Circuit copper orientation	face	1 = up	0 = down
Location	Circuit location on conveyor		1 = center	0 = edge
Airknife	State of air knives on flood module		1 = on	0 = off
Sqroller	State of squeegee rolls: flood exit		1 = in place	0 = out
Finrinse	State of final rinse		1 = on	0 = off
Thick	Thickness of Coating	microns		

duction, research and development, and quality assurance were asked to indicate variables that might affect the coating thickness and that should be included as potential variables in the experimental design. After considerable discussion, J. B. decided to include the variables listed in Table 26.1 as potential predictors of coating thickness. The thickness of the coating was measured in microns.

The experiment was carried out under closely controlled conditions over a set of twenty-three observations, each of which used various combinations of the potential influencing variables.[2] This set of observations was established using a standard linear design with center point. A series of 6-by-8-inch test samples were cut from single-sided, 1-ounce ED copper, laminated to 0.003-inch-thick polyester film. Samples were secured to the trailing edge of a rigid 12-by-18-by-0.06-inch piece of glass-epoxy laminate to ensure that the samples would move through the conveyor rollers at constant speed. The thickness of the Entek-plus protective film was measured via a process involving three steps:

1. Cut a 10-square-centimeter section from each sample.

2. Dissolve the coating from the surface in 25 milliters of 5 hydrochloric acid.

3. Measure the absorbency of that solution at a wavelength of 270 nanometers, using a UV spectrophotometer with a slit width of 1.0 nanometer. The coating thickness (in microns) was assumed to be equal to the absorbance.

The data gathered from the experiment are shown in Table 26.2 and are stored in a data file named Sheldahl.

[2]The data for this case were supplied by Sheldahl Inc. The assistance and cooperation of James Donaghy, president, Keith Casson, vice president for manufacturing, Jean Bronk, quality assurance manager, and James Munson, R & D engineer, are gratefully acknowledged.

TABLE 26.2 DATA FROM THE EXPERIMENTAL DESIGN

Trial	Web Speed	Bath Temperature	P1s Speed	P2 Speed	Orientation	Location	Air Knife	Sq Roller	Final Rinse	Thickness
1	10	100	5.0	5.0	1	1	1	1	1	0.549
1	10	100	5.0	5.0	1	1	1	1	1	0.352
2	60	120	50.0	50.0	0	0	0	0	0	0.307
2	60	120	50.0	50.0	0	0	0	0	0	0.305
3	10	100	5.0	5.0	1	0	0	0	0	0.504
3	10	100	5.0	5.0	1	1	0	0	0	0.954
4	60	120	50.0	50.0	0	0	1	1	1	0.200
4	60	120	50.0	50.0	0	1	1	1	1	0.272
5	10	100	5.0	50.0	0	1	1	0	0	0.533
5	10	100	5.0	50.0	0	1	1	0	0	0.541
6	60	120	50.0	5.0	1	0	0	1	1	0.277
7	10	100	50.0	50.0	0	0	0	1	1	0.490
8	60	120	50.0	5.0	1	1	1	0	0	0.298
9	10	100	50.0	5.0	0	1	1	0	0	0.542
10	60	120	5.0	50.0	1	0	1	1	1	0.275
11	60	100	50.0	50.0	1	0	1	1	0	0.251
12	60	120	5.0	5.0	0	0	1	1	1	0.285
13	10	120	50.0	5.0	0	0	1	0	1	0.525
14	60	100	5.0	5.0	0	1	0	0	0	0.232
15	10	120	5.0	50.0	1	1	0	1	1	0.772
16	10	100	50.0	50.0	1	1	1	0	0	0.452
17	35	110	27.5	27.5	1	1	1	1	1	0.258
18	35	110	27.5	27.5	0	0	0	0	0	0.426

You are to perform an analysis to identify which of the nine potential variables influence coating thickness. In addition, you are to present an estimated linear model whose coefficients indicate the effect of each variable on coating thickness. This model can subsequently be used to set the coating line to achieve the desired coating thickness at minimum cost.

Your analysis should include the following tasks.

26.1 Load the data from the Sheldahl file into your local computer.

26.2 Perform analyses to identify variables that influence thickness. Indicate the effect that each important variable has on coating thickness.

26.3 Prepare a written report presenting your analysis and discussing the results. Recommend appropriate settings for the production line.

NEW CONCEPTS FINANCIAL INC.

27.1 INTRODUCTION AND BACKGROUND

New Concepts Financial Inc. is a new small investment management firm organized in the last year by three former college friends. Their objective is to develop and market a dynamic investment strategy for small investors. To this end, they plan to utilize efficient computer information systems to reduce the transaction cost, to gather and catalog recent market data, and to rapidly analyze alternative investment portfolios.

The friends decided to organize this firm jointly because they recognize the unique contributions that each can make. Charlie Olson will develop and manage the marketing operations, drawing on his five years of experience developing investment products for a large bank and managing several products for a medium-size consumer foods company. Anita Lopez, who is in charge of systems development and operations, has four years of experience with a national computer systems development company, where her major assignments involved developing and installing systems for several New York investment houses. Bill Green, who will develop investment and portfolio strategies, has worked as an analyst and fund manager for a firm that specializes in retirement fund management.

During the past six months, the firm has attracted a growing number of clients through the efforts of Charlie and from contacts that Bill developed in his previous position. Payments from these clients provide enough income to cover basic operating costs, including modest salaries for the three principles. The firm continues to draw on the original loan, however, to develop computer systems, analysis procedures, and marketing efforts. The partners strongly believe that their business will ultimately be successful only if they develop effective computer-based information and analysis tools.

Anita and her staff of two have devoted most of their efforts to developing the system for obtaining daily market transactions and publicly available market analyses. Their productivity has been greatly increased by their use of purchased fourth-generation software to implement special applications. New Concepts also plans to purchase subscriptions to several research

TABLE 27.1	VARIABLES CONTAINED IN THE CONCEPTS DATA FILE
VARIABLE NAME	**VARIABLE DESCRIPTION**
Date	Date for the closing stock price
Tenseq	Numerical sequences; 1 to 10 for selecting 10-day intervals
Twenseq	Numerical sequences; 1 to 20 for selecting 20-day intervals
Mktindex	Closing S&P 500 Index for the given day
ABXprice	Closing price for stock ABX
BAprice	Closing price for stock BA Boeing
IBMprice	Closing price for stock IBM Inc
GLDprice	Closing price for stock GLD
KOprice	Closing price for stock KO Coca Cola
TXprice	Closing price for stock TX Texaco
UKprice	Closing price for stock UK Union Carbide
WXprice	Closing price for stock WX Westinghouse

service reports. But the company is still gathering information to help select appropriate research services. Anita has also developed software that enables the firm to buy and sell shares efficiently at the lowest possible cost. This software has recently been implemented, and the improvements are quite impressive.

27.2 PILOT STUDY ANALYSIS

Bill has developed preliminary designs for a number of potential strategies that he believes can provide value added for small investors. His interest in financial theory and practice was stimulated in graduate school by several Ph.D.-level courses he took in finance and in statistics. He was a co-author, with one of his professors, of a paper that appeared in a finance journal. Working from this background, he developed and revised several innovative products for the investment management firm.

Bill has hired you to develop pilot versions of analysis procedures for combining various stocks into portfolios with different levels of expected return and risk. These procedures are to begin with an analysis of recent stock price data and then use the analysis results to develop various portfolio options. In this first step, you will develop and implement the analysis by using a statistical analysis package and a computer spreadsheet. In the analysis model, closing prices for the stocks listed in Table 27.1 are used to compute percentage rates of return for intervals of ten days (two weeks) and twenty days (four weeks). Then the means and variances of the returns from each stock and the covariances between the returns from the various stocks will be computed. Bill has asked that, as you carry out the analysis, you compare the results from both intervals.[1]

Next, using these results, you are to compute the expected return and risk (measured by variance) for various stock portfolios. Portfolios would use various stocks combined in different proportions. For this initial analysis, portfolios with only three different stocks will be analyzed. After the procedure has been developed and shown to work, Anita's group will develop computer programs for automating your procedure and applying it to portfolios with an unlimited number of stocks. The plan is to install these programs on notebook computers, along with daily updated financial data. Financial planners employed by New Concepts will use the notebook computers to develop customized portfolios for their clients.

The procedure for computing expected returns and risk for various three-stock portfolios involves standard finance and statistical theory. This theory is presented in detail in finance and investment textbooks and is summarized in the case appendix here (section 27.3). Because the computations for each possible portfolio are time-consuming and subject to computational errors, a small spreadsheet is to be developed to perform the computations. The spreadsheet should enable the user to enter the means, variances, covariances, and names for each of three stocks. In addition, the procedure should allow the user to enter five different proportional combinations of the three stocks. From these data, the program will compute the expected return and variance for each of the five possible portfolios based on different percentages of the three stocks. By using this output, financial planners employed by New Concepts Inc. can provide clients with the combinations of return and risk which best meets their needs.

After discussing the situation with Bill Green, you agree to perform the following tasks for this analysis.

27.1 Select ten-day and twenty-day closing prices for each of the stocks in the Concept data file, and create new variable columns.

27.2 Compute the actual return percentage, based on the ten-day and twenty-day closing prices for each stock. Convert this return into an annual base, by multiplying the ten-day return by 26 and the twenty-day return by 13.

27.3 Prepare x-bar and R statistical process control charts, using samples of size $n = 3$ to study the stability of the rates of return over the one-year time period, based on the ten-day closing price intervals.

27.4 Compute the mean, variance, and covariances for the stocks for both the ten-day and the twenty-day returns. Present the means and the

---

[1]Special thanks to my faculty colleague Richard Goedde, who provided the initial data file.

variances for each stock in a table, and discuss the differences between the stocks in terms of return and risk. Include in your discussion a comparison between the ten-day and twenty-day return data.

27.5 Compute the correlation of each stock with the S&P 500 Index.

27.6 Prepare a spreadsheet model to compute expected return and variance for portfolios of three different stocks, in different combinations.

27.7 Construct two different three-stock portfolios. (Each stock considered in this case can be in only one portfolio.) Compare the mean return and risk for various percentages of the stocks in each portfolio.

27.8 Prepare a concise well-written report presenting the results of your analysis. Describe the model that you have developed, and emphasize its features. Discuss the expected returns and risks of the portfolios you have constructed, and indicate how the distribution parameters— mean, variance, and covariance—and the percentage of each stock contribute to the return and risk for different portfolios. Discuss any differences in results between using ten-day and using twenty-day closing price intervals. Indicate advantages and disadvantages of each procedure. An appendix to your report should document the procedure for computing returns from daily stock closing prices.

27.3 APPENDIX: BASIC THEORY FOR PORTFOLIO RISK AND RETURN

The theory required for this case if found in most good finance and investment textbooks, as well as in a number of business statistics textbooks. Financial analysts are concerned with the expected or mean return of a portfolio of stocks and with the risk for the particular portfolio. Everything else being equal, an investor would prefer a higher mean return. In general, investors demand a higher mean return if the risk is higher.

In finance, risk is directly related to the variance of the return from a portfolio. Larger variance implies a larger acceptance interval or confidence interval at a given probability of error, α. These larger intervals indicate a wider range of possible returns, implying a greater chance of obtaining a large positive return but also a greater risk of obtaining a small positive or even a negative return—and hence greater risk for the investor. A conservative investor would seek a stable value for his or her portfolio and thus a small-variance, small-risk portfolio. An investor seeking higher growth might seek a high-variance, high-risk portfolio. As indicated earlier, higher-risk portfolios require a higher mean return.

The statistical model for determining the expected return and the risk from portfolios is based on the linear combination of random variables, such as the form

$$W = \sum_{i=1}^{n} a_i X_i$$

where

W = Rate of return for the portfolio
a_i = Proportion of the portfolio for stock i
X_i = Rate of return for stock i
n = Number of stocks in the portfolio

The mean return for the stock portfolio, μ_w, is

$$\mu_w = \sum_{i=1}^{n} \mu_i$$

and the variance is

$$\sigma_w^2 = \sum_{i=1}^{n} \sum_{j=1}^{n} a_i a_j \sigma_{ij}$$

The estimators for the mean, μ_i, variance, σ_i, and covariance, σ_{ij}, obtained from an analysis of the rates of return for each stock are used in these equations to compute the mean and variance for various portfolios. When $i = j$, the covariance term σ_{ij} is the variance for the stock. To obtain lower-variance portfolios, an investor should choose stocks that have negative or small covariances between their returns. Negative covariances imply negative correlations, indicating that the returns from two negatively correlated stocks are moving in opposite directions and thus reducing the portfolio variance. Stocks with returns that are independent have zero covariance terms and thus their joint variations do not increase the portfolio variance. Stocks with positive correlations increase the portfolio variance because their changes move in the same direction—both up or both down.

CASE 28

BIG SKY POWER INC.

28.1 INTRODUCTION AND BACKGROUND

Big Sky Power Inc. is the electrical power company for the northern region of Loonland, an upper midwestern state. The company operates twelve electrical generating facilities: six coal-fired, four oil/natural gas-fired, one nuclear, and one wind farm. In addition, it buys and sells power from neighboring utilities to cover variations in demand pattern. To distribute this power, it operates and maintains an extensive distribution system consisting of substations and above-ground electrical lines, with newer residential lines tending to be built underground.

Because Big Sky has a monopoly over electrical power distribution in its region it is closely regulated by the Loonland State Public Service Commission. The regulation allows Big Sky to make a fair profit, provided that it operates efficiently. This arrangement requires considerable monitoring of Big Sky's operation by professional staff of the commission and annual hearings by the seven-member commission board, whose members are appointed to staggered six-year terms by the Governor of Loonland. A key decision each year involves the rates to be charged for electrical power to various residential, commercial, and industrial users. Big Sky proposes a rate per kilowatt-hour for each customer category, with supporting information concerning expected costs and revenues for the next year. Its proposals are reviewed by professional staff of the commission, who file their own analysis. Areas of disagreement are mediated in a formal hearing conducted by the commission, which then makes a final decision. The hearings often include technical material, including forecasts of power usage.

Each year Big Sky must present substantial evidence in support of its proposed rates. The company's total expected revenue is determined by multiplying the proposed rate for each category times the forecast unit sales of electrical power for each, and then summing these products. Total costs are composed of fixed costs plus variable generating costs per kilowatt-hour multiplied by the forecast unit sales of electrical power. The commission carefully monitors the claimed fixed costs to ensure that they are legitimate and do not include charges for past management errors. For example, charges to cover a cost overrun on a new power generation facility, which the commission ruled was due to management error, were not allowed. The nature of the business is that fixed costs are a major portion of total operating costs. Considerable debt service must be paid for generating facilities. Maintenance of distribution facilities is largely a function of the size of the facility and not of the amount of power distributed. Variable costs include fuel charges, purchased power, and connections for new users.

Clearly, the forecast of power usage for the next year is a critical component of the rate decision. The staff at Big Sky Power prepares the forecast, and the professional staff of the Public Service Commission reviews the forecast. If the forecast is too low and actual sales exceed the forecast, total revenue and total variable expenses will be larger than predicted. Because electric power distribution has large fixed costs, profits will be substantially larger than predicted if actual sales exceed forecast sales. Thus, Big Sky Power would like to have a low forecast, but the regulators would not. Conversely, a forecast that is too high translates into actual sales that fall below the forecast amount and, therefore, yield lower profits. Big Sky naturally desires to avoid optimistic forecasts. Because of the large fixed cost, differences between forecast and actual power consumption have considerable leverage on profits. Both Big Sky and the commission's professional staff desire an accurate forecast, but each side has a different preferred direction for the deviation between forecast and actual.

The Public Service Commission is also responsible for ensuring that sufficient electrical power exists to support future residential and business activity. Naturally Big Sky wishes to have sufficient generating capacity to provide for future demand and to avoid purchasing from neighboring power companies. Accurate long-run forecasts are very important because the lead time for new generating plants is six to eight years for planning, regulatory review, design, and construction. Therefore, in addition to wanting accurate short-term forecasts, Big Sky needs accurate long-term forecasts. In general, long-term forecasts are subject to considerable uncertainty because of major changes in the national and regional economy unanticipated by any rational observer. Thus, considerable experienced judgment is required.

28.2 FORECASTING METHODS

Analysis to develop time series forecasts begins with a plot of the series versus time. This plot is examined to obtain an initial understanding of the series. Initially the overall structure is examined. Does the series tend to increase or decrease? Is there evidence of cyclical and/or seasonal patterns?

Do the observations exhibit considerable deviation over the time interval? In some cases, the pattern and structure of the series can be inferred directly from the plot. But a number of sophisticated analysis tools can also be used to improve the forecast beyond the simple understanding obtained from examination of the plots.

There are two general strategies for forecasting time series data. Structural model strategies, which develop a relationship between a set of predictor variables and the time series; and time series model strategies, which use the past history of the series to forecast future values. A structural approach that uses multiple regression is workable if the variables that predict the series are well known. But to forecast it is necessary first to find or forecast values of the predictor variables in the multiple regression model. In contrast, a time series forecasting procedure identifies patterns in the series and/or relationships between lagged values to forecast future values. Thus you need not obtain measurements of other variables. Time series analysis procedures include components analysis, exponential smoothing, trend analysis, moving averages, and the Box/Jenkins autoregressive moving average. These approaches are developed in various textbooks, which should be consulted for details.

The choice between strategies depends in part on your understanding of the system that generates the series. If the system is well understood, a structural model for the time series can be specified (by a set of predictor variables and a mathematical form) and estimated. In many cases a linear model is a good approximation to system behavior, but other mathematical forms are often used. If the system is well understood and values of the predictor variables can easily be obtained for future periods, a structural approach is usually preferred. Within a structural model, you can adjust for specific factors, which makes the model more responsive to anticipated changes. Many analysts argue that, in contrast, systems are not well understood and that the choice of variables and model specification is arbitrary. In addition, a number of predictor variables must be forecast, and the potential errors in those forecasts complicate the forecast of the series being studied. Thus it may be preferable to analyze the data in the series and use the data to determine the forecasting model.

28.3   THE FORECASTING PROBLEM

Scott Fernandez, manager of household usage analysis, has asked you to prepare forecast models of residential electrical usage for use in future rate cases. Residential customers are grouped into those who use electricity for home heating and those who do not. Customers who use electricity for home heating are charged a lower rate per kilowatt-hour of electrical usage,

TABLE 28.1		VARIABLE NAMES IN THE BIGSKYRG DATA FILE
VARIABLE NAME	**NUMBER OF OBSERVATIONS**	**VARIABLE DESCRIPTION**
Salesmwh	69	Megawatt sales for residential regular customers, billing day and seasonally adjusted
Numcust	69	Number of residential regular customers, seasonally adjusted
Yd87	69	Personal disposable income (in 1987 dollars)
Pricelec	69	Price of electricity for residential customers (in 1987 dollars)
Aircnsat	69	Air conditioning saturation level
Degreday	69	Heating degree days base 50, departure from normal weather and billing day adjusted
Thi65	69	THI (temperature-humidity index) degree days base 65, departure from normal weather and billing day adjusted
Thi75	69	THI degree days base 75, departure from normal weather and billing day adjusted

because they purchase electricity in higher volumes. The data for electrical usage for regular residential customers is contained in a file named Bigskyrg and is described in Table 28.1. The data for electrical usage for residential customers with space heat is contained in a file named Bigskyht and is described in Table 28.2. You are to prepare forecasting models for each residential group.

In this project you will prepare forecasting models on the basis of quarterly data through and including 1993. Then you will test your forecasting model by applying it to the four quarters of 1994 and the first quarter of 1995. Forecasting models—especially those developed by using regression analysis—provide the best fit to the observations used to develop the models. After obtaining the best fit model using most of the available data, you can test this model on data not used to develop it. By using this procedure, you can determine how well the model will forecast when applied to unknown future time periods.

For this case, you will prepare a forecasting model for each of the two residential customer groups, using the two different data sets described below. Each data file contains duplicate copies of the variables. One set has a 2 included in the name of each variable. Variables with the 2 do not have the data from 1994 and the first quarter of 1995. You are to use these later variables to estimate the parameters of your regression and time series forecasting models. Then you will use each of those models to predict the data from 1994 and 1995. Finally you will compare the predicted and actual sales values to determine which of the forecasting models you developed provides the best forecast.

| TABLE 28.2 | VARIABLE NAMES IN THE BIGSKYHT DATA FILE |

VARIABLE NAME	NUMBER OF OBSERVATIONS	VARIABLE DESCRIPTION
Salesmwh	69	Megawatt sales for residential customers with space heat, billing day and seasonally adjusted
Numcust	69	Number of residential customers, with space heat, seasonally adjusted
Yd87	69	Personal disposable income (in 1987 dollars)
Pricelec	69	Price of electricity for residential customers with space heat (in 1987 dollars)
Aircnsat	69	Air conditioning saturation level
Degreday	69	Heating degree days base 50, departure from normal weather and billing day adjusted
Thi65	69	THI (temperature-humidity index) degree days base 65, departure from normal weather and billing day adjusted

28.1 Prepare time series plots for total electrical sales and other variables that are likely predictors of sales.

28.2 Prepare descriptive statistics and correlations for the variables in the data file. Analyze these to help determine important predictor variables for your forecast models.

28.3 Estimate the coefficients of a multiple regression model, using electricity sales as the dependent variable. After you have obtained your best model, use it to predict sales for 1994 and 1995. Compare the observed and forecast values for these test periods.

28.4 Use two different time series forecasting models to obtain forecasts for 1994 and 1995. Develop the model by using the data that exclude 1994 and 1995. Compare the observed and forecast values for these test periods.

28.5 Prepare a written report that presents the models and their forecasts. Based on your analysis, recommend a forecasting strategy for future electrical consumption.